Plant Genetic Manipulation for Crop Protection

Edited by

A.M.R. Gatehouse, V.A. Hilder and D. Boulter

Department of Biological Sciences
University of Durham
Science Laboratories
South Road
Durham DH1 3LE
United Kingdom

C·A·B International

C·A·B International   Tel: Wallingford (0491) 32111
Wallingford   Telex: 847964 (COMAGG G)
Oxon OX10 8DE   Telecom Gold/Dialcom: 84: CAU001
UK   Fax: (0491) 33508

A catalogue record for this book is available from the British Library

ISBN 0 85198 707 9
ISSN 0960-202X

Typeset by Leaper & Gard Ltd, Bristol
Printed and bound in the UK by Redwood Press Ltd, Melksham

Contents

Contributors

PETER D. BARFOOT   *Agricultural Genetics Company Ltd., 154 Science Park, Milton Road, Cambridge, CB4 4GG, UK.*

DONALD BOULTER   *Department of Biological Sciences, University of Durham, Science Laboratories, South Road, Durham, DH1 3LE, UK.*

RICHARD J.A. CONNETT   *Agricultural Genetics Company Ltd., 154 Science Park, Milton Road, Cambridge, CB4 4GG, UK.*

ANGHARAD M.R. GATEHOUSE   *Department of Biological Sciences, University of Durham, Science Laboratories, South Road, Durham, DH1 3LE, UK.*

DAVID L. HALLAHAN   *AFRC Institute of Arable Crops Research, Rothamsted Experimental Station, Harpenden, Hertfordshire, AL5 2JQ, UK.*

VAUGHAN A. HILDER   *(AGC Ltd) Department of Biological Sciences, University of Durham, Science Laboratories, South Road, Durham, DH1 3LE, UK.*

MICHAEL A. MAYO   *Scottish Crop Research Institute, Invergowrie, Dundee, DD2 5DA, UK.*

PHILLIP MORRIS   *AFRC Institute of Grassland and Environmental Research, Welsh Plant Breeding Station, Aberystwyth, Dyfed, SY23 3EB, UK.*

PHILIP M. MULLINEAUX   *John Innes Institute, John Innes Centre for Plant Science Research, Colney Lane, Norwich, NR4 7UH, UK.*

H. JOHN NEWBURY   *School of Biological Sciences, University of Birmingham, PO Box 363, Birmingham, B15 2TT, UK.*

MARNIX PEFEROEN   *Plant Genetic Systems N.V., J. Plateaustraat 22, 9000 Gent, Belgium.*

JOHN A. PICKETT   *AFRC Institute of Arable Crops Research, Rothamsted*

Experimental Station, Harpenden, Hertfordshire, AL5 2JQ, UK.

BRIAN REAVY   *Scottish Crop Research Institute, Invergowrie, Dundee, DD2 5DA, UK.*

LESTER J. WADHAMS   *AFRC Institute of Arable Crops Research, Rothamsted Experimental Station, Harpenden, Hertfordshire, AL5 2JQ, UK.*

ROGER M. WALLSGROVE   *AFRC Institute of Arable Crops Research, Rothamsted Experimental Station, Harpenden, Hertfordshire, AL5 2JQ, UK.*

K. JUDITH WEBB   *AFRC Institute of Grassland and Environmental Research, Welsh Plant Breeding Station, Aberystwyth, Dyfed, SY23 3EB, UK.*

CHRISTINE M. WOODCOCK   *AFRC Institute of Arable Crops Research, Rothamsted Experimental Station, Harpenden, Hertfordshire, AL5 2JQ, UK.*

HAROLD W. WOOLHOUSE   *Waite Agricultural Research Institute, University of Adelaide, Glen Osmond, South Australia 5064.*

Preface

It is becoming clear that the 1990s will be the decade in which genetically engineered plants move out from the research laboratory, and the containment glasshouse and plot, to the fields and greenhouses of farmers and commercial growers. However, this transition is unlikely to be entirely smooth. For various reasons, some well-founded, some not, the public in many developed countries has set its opinion firmly against transgenic crops; and the 1990s have already seen the first demonstration against the release of plants containing foreign genes, when an entirely innocent experiment to assess the frequency with which a transposable element moved in the plant genome was portrayed as an outrage against nature.

To some extent, the scientific community has itself to blame for the suspicion with which the public regards anything involving genetic engineering; there were far too many wild claims for the technology made during the 1980s, and far too few sober appraisals of what problems it was best suited to tackle. In the plant genetic engineering area, the early publicity surrounding the development of transgenic plants exhibiting herbicide resistance reinforced suspicions that this was only another way of making sure that the agrochemical industry maintained its control over farming practices. Ecological movements, which had become a political force to be reckoned with, were alienated from a technology whose incomprehensible jargon seemed to hide a threat to the environment worse than anything mechanized chemical agriculture had yet produced.

This breakdown in communication between the scientists and the general public is a serious problem, and plant genetic engineering has suffered as much from it as any other area of scientific advance. While it is possible that the attitudes of those opposed to transgenic technologies are now too rigid to admit of any real compromise, it is still worth stating the position of many plant scientists, including the editors of this volume. We are convinced that plant genetic engineering offers great benefits to the environment, by replacing the present policy of blanket sprayings of crops with herbicides, fungicides and pesticides (most of which never reach the organisms they are meant to affect) with a combination of inherent engineered resistance to pests and diseases, and selective treatments with specific 'safe' chemicals. Nor is this a technology to be limited to the developed countries; it is ideally suited to Third World agriculture, since it is 'user-friendly'. Unlike some of the 'improved' crop varieties that have been produced, the farmer can utilize the genetically engineered plant material in traditional agricultural practices, and is not forced to provide a chemical input in order for the crop to grow successfully. If it is applied in a sensible manner, there can be no doubt that this technology is 'green', even if it is difficult to convince Greens of its potential benefits.

Of course, plant genetic engineering cannot be excluded from regulations governing transgenic experimentation, since the technology must be proved to be safe before it can be applied in situations where the public may come into contact with it. Although anything expressed in a transgenic plant is confined to the tissues of that plant, and cannot contaminate the environment or affect organisms that do not utilize the transgenic plant as a host, the possibilities of transmission of engineered characters to other plants must be considered, as must the possibility of affecting the balance of ecosystems (although modern agricultural practices are almost as far removed from a balanced ecosystem as possible, and any amelioration of the chemical warfare strategy can only be favourable to the environment). However, the potential hazards inherent in transgenic plants are orders of magnitude less than those in transgenic bacteria or animals, nor is there any moral dilemma to be faced in the production of plants expressing foreign genes. We would therefore make a plea for greater tolerance towards this new technology as it moves out of the laboratory and into the marketplace, and would (humbly) suggest to environmental pressure groups that perhaps it is something they should be supporting – possibly even getting involved in – as a means of protecting the environment against applied chemicals.

The review chapters in this volume, written by scientists intimately involved with the development and application of plant genetic engineering, give a thorough coverage of the present 'state of the art' in transgenic crop protection technology. While the main audience for the volume will be specialists, we hope the general reader, who is perhaps interested in this area but less knowledgeable about it, may appreciate the potential and the fascination of plant genetic engineering.

The editors would like to take this opportunity of thanking all the contributing authors, and of acknowledging the assistance of Gordon's gin, Baby E., and the Bhagavad Gita in compiling this volume. Especially we would like to express our thanks to Dr John A. Gatehouse for his help in the compilation of this volume and for many useful discussions on the technology and applications of plant genetic engineering.

The Editors
Durham, 1991

Chapter 1
Introduction

Vaughan A. Hilder, Donald Boulter and Angharad M.R. Gatehouse

*Department of Biological Sciences, University of Durham,
South Road, Durham, DH1 3LE, UK*

Humankind faces a serious problem today. We have become totally dependent upon agricultural production to provide our food. Unfortunately, current agricultural output is unable to meet even the most basic needs of a sizeable fraction of humanity, with an estimated 450–1500 million people suffering from hunger and malnutrition (UN, 1988). It is far less able to satisfy the very reasonable aspirations of many people in the developing nations. This problem is compounded to the edge of catastrophe by the extent to which the increase in world population is outstripping increases in agricultural output.

This problem is most obvious in parts of the Third World, where it may amount to a matter of life or death. However, the sophisticated, high-tech agricultural systems of the developed world are subject to problems of their own. The high unit output of such systems is heavily dependent on high levels of industrial inputs. Such an approach is now widely seen as being unsustainable and is increasingly subject to criticism in terms of being:

1. extremely costly of non-renewable resources;
2. inefficient – inasmuch as a large proportion of the input chemicals miss their intended target and enter the environment; leading to
3. environmentally unacceptable – contamination of food chains and water sources, and the physical degradation of the environment is now widely recognized;
4. consumer unacceptable – concern over the problems outlined above is no longer confined to ecologists and other specialists but is becoming shared by a growing sector of the increasingly sophisticated public who question the indirect effects of high-input farming practices on themselves in terms of environmental damage and the direct effects in terms of residues in the produce which they consume.

1

The need is, therefore, for a massive increase in food production, amounting to around a 75% increase by the year 2000 (Blaxter, 1986), without a concomitant massive increase in dependence on non-renewable resources. The possibilities of meeting this need by increasing the area under cultivation, increased irrigation and improved cultivation practices are limited. A very significant contribution could, however, be made by protecting more of the crops which are currently grown from losses to pests, pathogens and weeds. Obtaining truly reliable figures for crop losses is notoriously difficult, but most estimates put the total loss of world-wide agricultural production at between 20 and 40%. These are primarily pre-harvest losses, although some post-harvest loss does also occur. Of this wasted resource, insect pests account for about 14%, plant diseases about 12% and weeds a further 10% (Figure 1.1). These losses occur despite the widespread use of synthetic pesticides, which had an end-user value of *c.* US$ 20 billion in 1987.

It is in this area of crop protection that the recent advances in biotechnology, especially in genetic engineering, have been looked to as offering the possibility of a revolutionary new solution. The methods by which transgenic plants can be produced are described in Chapter 2 of this volume. The list of major crop species which are amenable to such manipulation is growing steadily (Table 1.1), although there still remain some notable examples of major crops, particularly amongst the mono-cotyledons, which are holding out against the efforts to transform them.

Protection of crops from their insect pests was quickly seized upon as a major goal of plant genetic engineering. Insects not only are responsible for massive losses of productivity directly as a result of their herbivory, but also

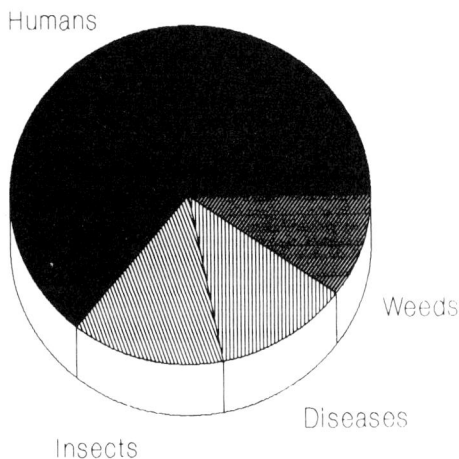

Figure 1.1. Division of the potential world agricultural output cake.

Table 1.1. Major world crops which are or are not yet amenable to genetic engineering. Figures in parentheses indicate the estimated world-wide value of the crop (billion US$) in 1967 (Cramer, 1967).

Transgenic plants produced	Unconfirmed reports of success	Transgenic plants not yet reported
Rice (36.4)		
		Wheat (24.3)
Maize (17.5)		
Potato (15.6)		
Cotton (11.8)		
		Sugarcane (9.3)
Grapes (8.6)		
	Cassava (7.1)	
	Sweet potato (5.3)	
Tobacco (5.2)		
Soyabean (4.8)		
Sugarbeet (4.6)		
		Coffee (3.7)
		Peanut (3.7)
	Common bean (3.2)	Citrus (3.2)
Apple (1.8)		
		Olive (1.6)
Tomato (1.5)		
		Pea (1.1)
Oilseed rape (1.0)		
Sunflower (1.0)		Cocoa (1.0)

serve as the vectors of many serious plant pathogens, and the physical damage they cause to the plant may facilitate infection by various other soil or airborne pathogenic organisms. Many quite different strategies are under investigation to exploit the opportunities afforded by genetic engineering to control insect pests. One type of approach is to manipulate the organisms which are naturally pathogens of the insects in order to enhance their effectiveness as biological control agents. *Bacillus thuringiensis* is a bacterial pathogen of certain insects which has been in limited field use as a biological control agent for more than 25 years. The genes which encode the insect control proteins responsible for its insecticidal activity have been cloned and transferred to other bacterial host species in an attempt to improve on its poor persistence in the field. Another recently reported example of this type of approach is the introduction of genes encoding the toxins from the venom of insectivorous spiders or mites into the viral genome of insect pathogenic baculoviruses in order to increase the rapidity with which the insects succumb to viral infection (Stewart *et al.*, 1991; Tomalski and Miller, 1991).

It is those cases where genes encoding insect control proteins have been introduced directly into the genome of a crop plant which are of direct relevance to this volume. Progress with transgenic plants expressing modified *B. thuringiensis* toxin genes is described in Chapter 6. Plants have evolved numerous mechanisms to protect themselves from insect attack, and some of these depend on insecticidal proteins. Work directed to moving these natural, plant-derived insect control protein genes into the genome of crop species is described in Chapter 7.

The reduction of crop losses due to plant diseases, both in terms of yield loss and reduction in food quality, is another area in which various types of biotechnology can make very major contributions. The identification and propagation of disease-free planting stock, based on immunological or nucleic acid hybridization techniques and plant tissue culture, have been important in eliminating, or at least reducing, viral pathogens from a number of crops. These techniques are also important in identifying the pathotype of various disease-causing organisms. In order to exploit the potential of genetic engineering to produce cultivars with inherent resistance to their major pathogens it is necessary to identify 'resistance genes' and, preferably, to develop some understanding of the mechanisms of the interaction between the resistance factor and target pathogen. Work directed to this end concerning fungal pathogens is described in Chapter 5. In the case of (some) viral pathogens, a number of alternative strategies have already been identified and exploited, as described and compared in Chapter 8.

Production of transgenic crops with resistance to, or tolerance of, herbicides is seen as a rather more controversial application of this technology. Unlike the examples of insect resistance and disease resistance this is not a substitution technology, but rather may increase the dependence on exogenously applied synthetic chemicals. There are arguments which can be advanced to support the suggestion that the selective use of herbicides and herbicide-resistant cultivars may be a major improvement on current practice for controlling weeds in some situations. Some of the issues involved are returned to in Chapter 10. There can be no doubt that, whatever the eventual usage in the field of such plants, the work to produce herbicide-tolerant transgenic plants has been central to the development of the whole of this technology. The various strategies by which transgenic plants may be made tolerant of herbicides are reviewed in Chapter 4.

It is still early days in the development of this technology and most of the work to date has concentrated on the introduction of a single gene which confers the desired resistance trait. Many of the natural defence mechanisms in plants involve interactions of numerous gene products in complex, multi-step metabolic pathways which give rise to small, non-protein secondary metabolites. The range of such pathways within the plant kingdom is vast. The manipulation of these secondary metabolic pathways in a regulated manner is clearly a much more difficult task than the introduction of a single new gene. The potential for such manipulation for genetically engineering resistance in crops in the future is considered in Chapter 9.

The use of genetically engineered crops is an entirely new and unproved technology. The scientific community, the legislators and the general public have had their fingers burned sufficiently often now by the wholesale adoption of new technologies before the wider consequences have been adequately examined. The movement of transgenic crop plants from the laboratory to the farmer's field is being advanced by a cautious, step-by-step approach. Some of the complex regulatory issues which are evolving to meet this new contingency are discussed in Chapter 3. This chapter also highlights the significant involvement of the commercial sector in the development of this technology and the factors which are likely to determine its translation into new commercial varieties of crop plants.

The realities of the situation of a crop in the farmer's field are, of course, far more complex than the experimental systems which have so far been studied. Chapter 10 points to some of the crucial issues which will have to be addressed if this technology is ultimately to be reduced to practice.

Since the first reports of transgenic plants appeared in 1984 (Horsch *et al.*, 1984) there has been exceptionally rapid progress directed at using this new technology for the very real, practical ends of crop improvement. It is to be hoped that the highly encouraging progress described in the subsequent chapters is maintained and developed to make a significant contribution towards redressing the imbalance between world food production and requirements. Something has to.

References

Blaxter, K. (1986) *People, Food and Resources.* Cambridge University Press, Cambridge.

Cramer, H.H. (1967) *Plant Protection and World Crop Production.* Bayer Pflanzenschutz, Leverkusen.

Horsch, R.B., Fraley, R.T., Rogers, S.G., Sanders, P.R., Lloyd, A. and Hoffman, H. (1984) Inheritance of functional foreign genes in plants. *Science* 223, 496–498.

Stewart, L.M.D., Hirst, M., Ferber, M.L., Merryweather, A.T., Cayley, P.J. and Possee, R.D. (1991) Construction of an improved baculovirus insecticide containing an insect-specific toxin gene. *Nature* 352, 85–88.

Tomalski, M.D. and Miller, L.K. (1991) Insect paralysis by baculovirus-mediated expression of a mite neurotoxin gene. *Nature* 352, 82–85.

UN World Food Council (1988) *The Global State of Hunger and Malnutrition: 1988 Report.* 14th Ministerial Session, Nicosia, Cyprus.

Chapter 2
Methodologies of Plant Transformation

K. Judith Webb and Phillip Morris

AFRC Institute for Grassland and Environmental Research,
Welsh Plant Breeding Station, Aberystwyth, Dyfed,
SY23 3EB, UK

Introduction

Genetic transformation can be defined as the transfer of foreign genes isolated from plants, viruses, bacteria or animals into a new genetic background. In plants, successful genetic transformation requires the production of normal, fertile plants which express the newly inserted genes. Transformation can utilize gametophytic tissues, prior to fertilization, or somatic cells, such as explants, cells and protoplasts, which can be stimulated to regenerate plants *de novo* in culture. In the latter case, such techniques must be available, not only for the species in question, but also for the specific cultivars, genotypes and tissues.

The process of genetic transformation involves several distinct stages, namely insertion, integration, expression and inheritance of the new DNA. Methods of gene insertion can involve the use of bacterial (*Agrobacterium* species) or viral vectors and direct gene transfer (DGT). These techniques utilize similar gene constructs, comprising bacterial or viral promoters linked to appropriate genes. The method of gene transfer which is chosen is generally governed by the plant species and its regenerative response in tissue culture. Both vectors and DGT have their part to play in the transformation of monocotyledonous and dicotyledonous species, although one approach may be more appropriate than another in certain instances.

In many systems, transient expression of novel genes in cells has allowed refinement of the methods of insertion, but ultimately stable integration of the DNA into the plant's genome is required. The continued expression and stability of the novel DNA in succeeding generations is also essential, particularly in seed crops.

Here, we will outline progress made in the identification of cells and tissues suitable for transformation, the construction of the genes used to monitor successful transformation, the methods of gene transfer and, where appropriate, the expression and inheritance of the foreign genes in the resulting transgenic plants.

1. Target cells for genetic transformation

A basic requirement for successful genetic transformation is the ability of the target cells, once transformed, to develop into complete plants. Such cells are found in embryos, ovules and pollen grains. Techniques of tissue culture extend the potential target cells to include those found in plant organs, such as leaves, stems, hypocotyls and cotyledons, and in callus or suspension cultures, and even isolated protoplasts. These cells can be stimulated to regenerate into new plants, via initiation of shoots *de novo* or via embryogenesis.

The ability of cells to regenerate into plants in culture varies not only with the cells and tissues themselves but also with the species, cultivar and even the individual genotype. A major difference in ease of regeneration is found between monocotyledonous and dicotyledonous species: monocots are generally considered less amenable in culture than dicots. Leaves of many dicot species can be directly transformed via *Agrobacterium* or used as a source of protoplasts for direct gene transfer. In monocots, immature or juvenile tissues provide the best source of potentially embryonic cells. These cells must first be multiplied to give cell cultures, which are then used both as a source of embryogenic cell clusters for gene transfer by microinjection or particle bombardment and as a source of protoplasts for DNA transfer by chemical or electrical means.

Ideally, all plants regenerated from cells and tissues should be identical to the parental plant. But the very process of plant regeneration can disrupt the genetic stability of the cells, creating genetic variation. Such changes are referred to as somaclonal variation (Scowcroft and Larkin, 1982). This variation can exist as gross cytological changes, such as translocations deletions or inversions, or more subtle changes, such as point mutations and rearrangements, and may itself be useful in creating novel variants (Evans, 1989; Lal and Lal, 1990). In a genetic transformation system, the ultimate aim is to regenerate plants identical to the parental material, except for the newly inserted genes.

Various factors influence somaclonal variation. The choice of tissue for culture, the constituents of the media and the duration of culture are all known to affect genetic stability of regenerants. Less variation exists in plants which have regenerated from pre-existing meristems, such as shoot or root meristems, than in those produced from the less organized suspension cultures or isolated protoplasts (Stafford, 1991). Therefore, regenerants must be closely monitored to establish the expected frequency and types of somaclonal variants created by a given regeneration system and, where possible, the amount of variation should be minimized.

Tissues from a range of species, including a number of the most important agricultural crops, can now be regenerated into plants, by the production of shoots or embryoids (Lal and Lal, 1990). Some of the most notable successes will be referred to in the following sections.

2. Genes and their construction

2.1 Basic organization of plant genes

Nuclear plant genes consist of distinct regions, each with different functions, involved in transcription and translation of mRNA (Figure 2.1). Starting with the 5′ end, these are: a promoter region which is involved in the initiation of transcription, together with enhancer/silencer regions which confer environmental or developmental regulation of expression (for example, light regulated regions), a transcriptional start or cap site and the so-called CAAT and TATA boxes which are located upstream from the cap site and which probably function in RNA polymerase II binding. Following the cap site is an untranslated region preceding the translation initiation codon (ATG). Within the transcribed region one or more untranslated or intron regions may be present. The end of the translated region is determined by a stop (TAA, TAG or TGA) codon, followed by

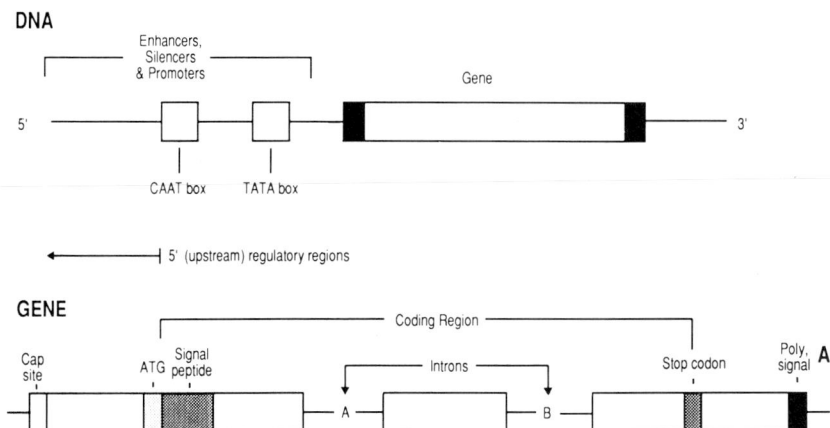

Figure 2.1. Basic structure of plant genes.

an untranslated region and finally terminated at the 3′ end by a poly-adenylation signal (typically G/AATAA $(A)_n$).

In addition, where proteins have to cross membranes (for example, into chloroplasts or vacuoles) or be inserted into membranes, their genes contain 5′ additions which encode transit or signal peptides respectively. These peptides are later cleaved to produce the mature proteins.

2.2 Promoters and other regulatory DNA sequences

For expression in plant cells foreign genes need to have appropriate promoter, 5′ leader and 3′ terminator sequences to ensure efficient transcription, stability and translation of mRNA. The large differences between regulatory elements from prokaryotes and eukaryotes mean that these sequences from the bacteria may not function in plant cells. Exceptions to this are the regulatory elements of certain genes of *Agrobacterium tumefaciens* and *A. rhizogenes* which are active on transfer to plant cells. Thus the *nos* (nopaline synthase), *ocs* (octopine synthase) and *mas* (mannopine synthase) gene promoters have been used success-fully to direct expression of genes in plant cells. In addition, plant viruses which depend upon plant transcription and translation factors have been used as sources of regulatory elements. The most common of these is the promoter of the *35S* RNA gene of the cauliflower mosaic virus (CaMV). This promoter is active in all tissues but its activity varies between different cell types. It is now the most widely used constitutive promoter for both dicot and monocot transformation, giving 10–40-fold higher levels of expression than either the *19S* CaMV or the *nos* promoter from *A. tumefaciens.*

Growing evidence of less efficient functioning of the *35S* promoter in monocot cells (at least 10–100-fold lower) compared with dicot cells has resulted in a search for a more efficient constitutive promoter for monocots. One development arising from the finding that the *35S* promoter consists of several domains, which confer different develop-mental and tissue-specific expression patterns, is that doubling of certain enhancer regions of the promoter increases overall activity by several fold and confers higher levels of expression in both dicot and monocot systems. A further development is the use of the promoter derived from the maize alcohol dehydrogenase 1 (*Adh1*) 5′ flanking sequence, which shows anaerobic induction and levels of expression in monocot cells equivalent to, or higher than, CaMV *35S* promoter. Transgenic rice plants expressing *gus* under the *Adh1* promoter, obtained via protoplast transformation with polyethylene glycol (PEG), showed expression primarily in roots indicating correct regulation of this gene in other monocot systems (Zhang and Wu, 1988).

The presence of the *Adh1* intron 1 between the promoter and coding region has also been shown to increase expression of the *nptII* gene in maize (Callis *et al.*, 1987). When the intron was used with the CaMV *35S* promoter, higher levels of expression were also found in barley and wheat protoplasts than were obtained without the intron region. Similar results were obtained with the first intron of the maize shrunken-1 locus. This increased transient expression of *gus* up to 90-fold in electroporated protoplasts of *Panicum maximum, Pennisetum purpurea* and *Zea mays* when driven by the CaMV *35S* promoter compared with plasmid constructs without the promoter, and up to 10-fold higher levels of expression than with the *Adh1* intron (Vasil *et al.*, 1989).

2.3 The use of reporter genes to study gene expression

The expression of a foreign gene by transformed plant cells can best be assessed by determining the abundance or activity of its product. The coding sequences of genes for bacterial enzymes, which are easily assayed and whose activity is not normally found in plants, form the basis of many reporter genes. Commonly used enzymes include bacterial nopaline synthase (*nos*) and chloramphenicol acetyltransferase (*cat*), bacterial or firefly luciferase (*lux*) and bacterial neomycin phosphotransferase (*nptII*) and β-glucuronidase (*gus*). These reporter genes (either as transcriptional or, more rarely, translational fusions) have been used extensively to analyse the function of promoters and other gene regulatory sequences. The availability of these reporter genes, which have sensitive, convenient and reliable enzymatic assays, has greatly increased the usefulness of transient assays (section 6.1).

The various reporter genes have several advantages and disadvantages in use and should be chosen carefully for particular applications. For example, the luciferase enzyme has the advantage that there is little or no endogenous luciferase activity in plants whereas endogenous enzymes in some plants are able to acetylate chloramphenicol and hydrolyse *gus* substrates. A considerable advantage with both *lux* and *gus* genes is that they can be used as visual markers in cells and tissues to determine the distribution of gene expression. In fact, the assay for *gus* activity is so sensitive that it can be used to visualize single cells which express *gus*, using a histochemical assay. It has therefore been widely used to determine the frequency, type and distribution of cells expressing activity after introduction of DNA by microprojectile bombardment (with the DNA gun) and after gene transfer to protoplasts. However, a disadvantage for applications relating to studies of gene expression is the high stability of the *gus* enzyme, which can persist for days or weeks after transcription and translation have ceased. Similarly, while luciferase activity has been demonstrated from the lux gene in carrot protoplasts only 8 hours after transformation, activity was still detectable 8 days later. However, the lux gene is particularly useful for tissue and whole-plant studies as luciferin is taken up into vascular tissue and luciferase activity can be simply detected on photographic film.

The nptII gene, on the other hand, has a much less sensitive and convenient assay, but it has the advantage of being a selectable marker in many plants. A comparison of nptII and cat genes as reporter and selectable marker genes found cat selection to be difficult, particularly in some Brassica species which contain high levels of endogenous cat activity. Therefore, cat should probably be used only as a reporter gene. Cat has been used extensively as a reporter gene in both dicot and monocot systems for comparing promoter activities and for optimizing electroporation conditions and could be detected within 30 minutes of polyethylene glycol (PEG)-induced uptake into tobacco mesophyll protoplasts, with maximum levels 4–24 hours post-transformation. Comparisons of cat, gus and nptII as reporter genes under the CaMV 35S promoter in tobacco showed that expression of the gus gene was the easiest to detect, followed by nptII and then cat. Recent results from several laboratories suggest that many species have endogenous gus-like activity: however, the pH optima of this activity are usually lower than for gus from E. coli. However, this endogenous activity can give high and variable background levels, particularly in prolonged histochemical studies.

2.4 Selectable marker genes

Selection of transformed cells is a key factor in developing successful methods for genetic transformation. Tumorigenic genes associated with pathological infection by Agrobacterium tumefaciens (Márton et al., 1979; Hernalsteens et al., 1980) and A. rhizogenes (Tempé and Casse-Delbart, 1989) can be used as selectable markers. These genes have profound effects on the morphology of the transformed tissues. In A. tumefaciens, crowngalls continue to grow, when cultured in vitro, in the absence of added plant hormones. Some strains of A. tumefaciens cause transformed cells to form abnormal shoots, which are generally infertile. The hairy root phenotype produced by cells infected by A. rhizogenes is equally distinctive but less disruptive and, in some species, also produces relatively normal, fertile plants.

Other characteristics, usually governed by single dominant genes encoding suitable resistance to a selective agent, can either replace (in A. tumefaciens) or supplement (in A. rhizogenes) the tumorigenic genes. These genes do not disrupt plant regeneration, but allow selection of transformants. These same selectable marker genes are also suitable for the identification of cells following direct gene transfer.

Several factors affect the efficacy of chemicals used for selection. The selection agent must be toxic to plant cells, though not so toxic that products from the dying, non-transformed cells kill adjacent, transformed cells. Thus the most effective toxins are those which either inhibit growth or slowly kill the non-transformed cells. Optimal selection pressure will use the lowest level of toxin needed to kill untransformed tissues.

Careful timing of the application of a selection pressure is critical in order to limit the number of non-transformed cells which survive through cross-protection by transformed cells. Genes conferring resistance to a variety of toxic compounds, such as methotrexate, antibiotics and broad-range herbicides, have been fused to suitable plant promoters (section 2.2) and used to select and identify transformed cells. Thus a range of selectable marker genes is now available for both *Agrobacterium*-mediated and direct gene transfer.

Methotrexate

Methotrexate has not been widely used although it was one of the first selection agents tested in dicotyledonous plants (Herrera-Estrella *et al.*, 1983). It is a cytotoxic drug which inhibits dihydrofolate reductase (*dhfr*), resulting in a deficiency of thymidylate, and ultimately death. A mutant *dhfr* gene, isolated from transposon Tn7 or plasmid R67, produces an enzyme which is resistant to methotrexate. Such a gene has also been successfully used to select for transformed callus in the monocotyledonous species, *Panicum maximum* (Hauptmann *et al.*, 1988).

Antibiotics

The antibiotics kanamycin, G418 and hygromycin are currently amongst the most widely used selection agents. All three are aminoglycosidic antibiotics which affect translational activities of the cells.

Neomycin phosphotransferase II (*nptII*) detoxifies, by phosphorylation, neomycin, kanamycin and G418. The *nptII* gene from transposon Tn5 has been widely used in dicotyledonous systems, including tobacco, potato and tomato (An *et al.*, 1986), legumes, such as white clover (White and Greenwood, 1987) and pea (Puonti-Kaerlas *et al.*, 1989), and woody species, such as *Pseudostuga menziesii* (Ellis *et al.*, 1989). But this marker gene has proved unsuitable for some dicotyledonous species, such as *Arabidopsis*, in which large numbers of non-transformed cells survive, and for many monocotyledonous species whose growth is not significantly inhibited by the antibiotic (Potrykus *et al.*, 1985; Hauptmann *et al.*, 1988; Dekeyser *et al.*, 1989).

Hygromycin phosphotransferase (*hptIV*) governs resistance to hygromycin. This gene, isolated from *E. coli*, has been placed under various promoters and has been successfully used in strawberry (Nehra *et al.*, 1990) and rice (Dekeyser *et al.*, 1989; Shimamoto *et al.*, 1989). Variable levels of resistance to hygromycin have been found in other graminaceous species (Hauptmann *et al.*, 1988).

Herbicides

The introduction of resistance to herbicides into agriculturally important crop plants can be an end in itself (Oxtoby and Hughes, 1990). Alternatively, the introduced gene can be used as a selectable marker gene. For example, glyphosate is the active ingredient in the broad-range herbicide, Roundup (Monsanto), which inhibits the enzyme, 5-enolpyruvyl shikimate-3-phosphate synthase (EPSPS). Modified versions of EPSPS, isolated from *E. coli* or *Salmonella typhimurium*, confer resistance to glyphosate on transformed plants. Such transformants have been produced in tobacco (Comai *et al.*, 1985), tomato (Filatti *et al.*, 1987), flax (Jordan and McHughen, 1988) and soyabean (Hinchee *et al.*, 1988).

Another, more recently developed system uses the *bar* gene isolated from *Streptomyces hygroscopicus*, which confers resistance to the herbicide, phosphinothricin (PTT), the active ingredient of Bialaphos and Basta (Hoechst). The *bar* gene codes for phosphinothricin acetyltransferase (PAT), which inactivates PTT, an irreversible inhibitor of glutamine synthetase. This gene has been inserted and expressed in several crop plants, including tobacco, tomato, potato (De Block *et al.*, 1987), rape (De Block *et al.*, 1989), alfalfa (Krieg *et al.*, 1990), maize (Spencer *et al.*, 1990) and rice (Dekeyser *et al.*, 1989).

Comparisons of these different selection agents under the control of different promoters has allowed optimal systems to be developed for many species. For example, working with rice, one of the better-established monocot systems, Dekeyser and co-workers (1989) evaluated the efficiency of various selectable markers for transformation. They found that, while methotrexate, phosphinothricin and bleomycin are effective at low concentrations, G418 and hygromycin are required at higher concentrations and kanamycin by contrast only partially inhibited growth of the rice cells.

3. Genetic transformation by *Agrobacterium* species

3.1 *Agrobacterium* species as plant pathogens

Crowngall and hairy root diseases are caused by *A. tumefaciens* and *A. rhizogenes* respectively. These phytopathogenic, Gram-negative bacteria (family Rhizobiaceae) are found in abundance in the soil and infect plant cells near wounds, usually at the crown of the root at the soil surface. Although the phytopathogenic properties of these bacteria were recognized over 80 years ago (Smith and Townsend, 1907), the true nature of the infection has been revealed relatively recently (Hooykaas and Schilperoort, 1984).

The molecular biology of infection by *A. tumefaciens* is better understood than that of *A. rhizogenes* but there are many similarities between the two species. Both are natural gene vectors. During infection, the bacterium transfers a small section of its own genetic material (T-DNA) into the genome of the host plant's cell. Once inserted, the bacterial genes are expressed by infected cells of that plant. By understanding and manipulating this process of infection or transformation, scientists have been able to harness these powerful and sophisticated vectors to transfer specific, cloned genes of major importance into dicotyledonous plants.

Characteristics of cells infected by Agrobacterium *species*

Crowngalls of *A. tumefaciens* show various morphological characteristics, ranging from disorganized masses of cells to the production of roots and shoots. By contrast, *A. rhizogenes* usually produces 'hairy roots' which are capable of growing in a negatively geotropic fashion. In some species, these hairy roots produce shoots and, ultimately, plants.

However, crowngalls and hairy roots share certain characteristics, such as the ability to grow in culture without the addition of plant hormones and the production of unusual amino acids, such as opines (Tempé and Casse-Delbart, 1989). These opines are utilized by the bacterial population as its source of carbon and nitrogen. The type of opine produced by crowngalls or hairy roots is determined by the bacterial strain, or more precisely by the particular plasmid carried by the strain, and not by the host plant.

Strains of *A. tumefaciens* are predominantly octopine or nopaline producers, whereas those from *A. rhizogenes* produce mainly agropine or mannopine. The detection of these distinctive opines in tumorigenic tissues has been widely used to confirm transformation, although opines are not always produced in successful transformations (Hernalsteens *et al.*, 1980). The genes for production of opines and phytohormones reside on large bacterial plasmids. These plasmids also carry genes which catabolize opines and govern the virulence and host range of the bacterium.

Structure of plasmids

Both bacteria contain large plasmids about 200 kbp (kilobase pairs) in size. *A. tumefaciens* carries the tumour-inducing or Ti plasmid whereas *A. rhizogenes* contains the root-inducing or Ri plasmid. These plasmids can be interchanged between the two species, or lost altogether if the bacteria are grown at high temperatures (above 28 °C), so producing avirulent strains of bacteria.

Both Ti and Ri plasmids can be conveniently considered in two component parts, only one of which, the T-DNA, is mobilized during infection. The genes for opine metabolism and phytohormone independence reside on this T-DNA, while genes governing virulence (*vir* genes) and opine catabolism reside outside this region.

Chemicals, such as acetosyringone, are produced by wounded plant cells. Over 8–16 hours, these exudates stimulate transcription of *vir* genes on the bacterial plasmid (Hooykaas and Schilperoort, 1984). These genes, in turn, produce endonucleases which excise a small fragment of DNA, the T-DNA, which traverses the cell wall and cytoplasm and finally integrates into the plant cell's genome.

Plasmid T-DNA

T-DNA is a small section of the plasmid DNA, about 23 kb in size, which makes up about 10% of the Ti or Ri plasmids. This stretch of DNA is flanked by 25 bp (base pair) repeated sequences which are recognized by the endonucleases encoded by the *vir* genes. Excision of the T-DNA initially produces a single-stranded molecule which is presumed to be an intermediate in the transfer to the cells (Zambryski *et al.*, 1989). The precise mechanisms of transfer of the T-DNA is unclear but may involve the production of circles of DNA (Koukolíková-Nicola *et al.*, 1985).

Some of the eukaryotic genes from T-DNA of the Ti and Ri plasmids have functions in common. For example, some genes (see below) confer phytohormone independence on the galls while others produce opines: neither of these two sets of genes is required for transfer of T-DNA or for stable maintenance of genes in the plant's genome. But here the similarities end. The differences between T-DNA from *A. tumefaciens* and *A. rhizogenes* have important implications for their use as gene vectors.

T-DNA OF *A. TUMEFACIENS*

Within the T-DNA, two distinct regions have been identified, the TL and TR regions. The T-DNA of nopaline strains can integrate as a single segment, whereas octopine strains frequently integrate as two segments, TL and TR. The former segment (TL) carries the genes controlling auxin and cytokinin biosynthesis and is always present when tumours are formed. Failure of the latter segment (TR) to integrate results in the loss of opine biosynthesis.

Auxin and cytokinin biosynthesis is controlled by three genes. Tms1 (iaaM) and tms2 (iaaH) code for enzymes of indole acetic acid biosynthesis while Tmr4 (ipt) codes for an enzyme in the synthesis of the cytokinin, isopentenyl adenine. These genes allow crowngalls to continue growth in culture in the absence of added plant hormones, but they also often inhibit the production of shoots and therefore regeneration.

T-DNA OF *A. RHIZOGENES*

As in *A. tumefaciens*, the T-DNA comprises two parts, the TL and TR regions. The TL-DNA codes for a set of genes, root locus or *rol* genes, A, B, C (White *et al.*, 1985; Spena *et al.*, 1987; Schmülling *et al.*, 1988; Capone *et al.*, 1989), which increase auxin sensitivity, rather than coding for auxin or cytokinin biosynthesis (Schmülling *et al.*, 1988). These genes from TL-DNA are always found in transformed roots but those from TR-DNA, encoding for opine biosynthesis and auxin biosynthesis (Tempé and Casse-Delbert, 1989), may not be present.

3.2 Transfer of genes of interest

Two strategies for gene transfer exist which utilize these natural gene vectors.

In *A. tumefaciens*, removal of the oncogenes from the T-DNA of the Ti plasmid permits the use of this bacterium for genetic transformation. In the presence of suitable media with auxins and cytokinins, normal but transgenic shoots and plants can be regenerated. Foreign genes can be placed within the boundaries of the T-DNA, either on the original virulent, but disarmed, Ti plasmid (*cis*) or on an additional, separate, non-virulent plasmid (*trans*).

Cis *or co-integrate vectors*

The novel, foreign genes, such as *nptII*, replace the oncogenic genes between the T-DNA border fragments on the virulent Ti plasmid. This foreign DNA which resides between the T-DNA borders is then transferred to the cell's genome. The limited space available for the new genes in these vectors has restricted the size and number of genes that can be transferred.

Trans *vectors*

An extra, non-virulent plasmid is transferred into the *A. tumefaciens* strain where a separate disarmed Ti plasmid already exists with a virulence region. Thus, transfer of T-DNA is mediated by activity of the *vir* region from one plasmid acting on another plasmid.

In *A. rhizogenes*, the hairy root phenotype itself can be used to identify the transformed cells. These bacterial strains are not disarmed, as has been described for *A. tumefaciens*, but similar strategies for gene transfer are used. The foreign genes are either inserted alongside the *rol* genes (*cis*) or into an additional, non-virulent plasmid (*trans*).

In the former case, the genes of interest will always be transferred to the same site as the *rol* genes which govern the hairy root phenotype, and cannot be separated from them in subsequent generations. In the latter case, three possibilities exist: the foreign genes alone are transferred, the *rol* genes alone are transferred or both the *rol* genes and the foreign genes are transferred simultaneously. The hairy root phenotype provides a visual marker for transformation and selectable marker genes allow selection of co-transformants. Since the co-transferred genes may insert into different sites and different chromosomes, these genes may then be separated from each other in subsequent generations.

4. Transformation of monocots by *Agrobacterium* species and viruses

The precision with which *Agrobacterium* transfers its DNA into the host plant's genome makes it an ideal method for genetic transformation. However, as state earlier, the apparent failure of *Agrobacterium* to infect monocotyledonous plants has restricted its application to the more amenable dicotyledonous species.

Recent work has clearly demonstrated that T-DNA can be incorporated into monocotyledonous species. Visible symptoms can be seen in non-graminaceous members of the monocotyledons. Small galls or swellings were seen at or near the sites of infection of some species and opines were detected in extracts from crowngalls of *Asparagus* grown in culture (Hernalsteens *et al.*, 1984) and from swellings in *Dioscorea bulbifera* (Schäfer *et al.*, 1987), *Chlorophytum capense* and *Narcissus* cv Paperwhite (Hooykaas-Van Slogteren *et al.*, 1984). More recently, Dommisse and co-workers (1990) reported that both *A. tumefaciens* and *A. rhizogenes* successfully infected *Allium cepa* and that opines were produced from transformed tissues. Opines have also been detected in graminaceous species such as *Zea mays* (Graves and Goldman, 1986). Direct evidence of the insertion of T-DNA, using molecular techniques has been obtained in *Dioscorea bulbifera* (Schäfer *et al.*, 1987) and in *Oryza sativa* (Rainer *et al.*, 1990). In the latter case, transformed cells produced callus which was resistant to kanamycin and which expressed the *gus* gene. The callus failed to generate any shoots.

The T-DNA from *Agrobacterium* can therefore be integrated into and expressed by the monocot genome. Further refinements of methods of DNA delivery by *Agrobacterium* species require sophisticated methods to permit rapid identification of successful transformation. Agroinfection magnifies such events, permitting even a single transformation event to be visualized as symptoms of viral disease in the whole plant.

4.1 Agroinfection with gemini viruses

Maize and wheat have been successfully transformed with the gemini virus genome, such as maize streak virus (MSV) and wheat dwarf virus (WDV), using *Agrobacterium* as a vector. Naked and cloned viral DNA of MSV failed to infect plants; the intact virus is infectious only when transmitted by the leafhopper insect (Grimsley *et al.*, 1987). However, the viral genome inserted between the T-DNA borders of *A. tumefaciens* did cause infection in the agriculturally important cereal crops, maize (Grimsley *et al.*, 1987) and wheat (Woolston *et al.*, 1988; Dale *et al.*, 1989; Marks *et al.*, 1989). Marks *et al.* (1989) demonstrated using wheat seedlings grown *in vitro* that the rate of infection with WDV varied not only with the strain of *Agrobacterium* but also with the bacterial species. Nopaline-producing strains of *A. tumefaciens* were more infective than octopine-producing ones but both were less infective than *A. rhizogenes.*

While embryos, seedlings and whole plants of different species have been used with varying degrees of success, transformation of cells capable of regenerating from suspension cultures would be an ideal method for producing transgenic plants. Experiments show that the bacteria can attach to cell walls of asparagus (Draper *et al.*, 1982), bamboo (Douglas *et al.*, 1985), maize, wheat and gladiolus (Graves *et al.*, 1988). These approaches, combined with the refinement of the delivery system, could well result in increased rates of infection and integration.

4.2 Transformation with viruses

Viruses offer some advantages as genetic vectors since they infect a wide range of species, including monocotyledonous species. In some infections, they also spread systemically throughout the plant. There is, however, no evidence to suggest that the genetic material introduced in this way can be permanently inserted into the plant genome, and so be heritable.

Viruses mainly have genomes of RNA, but the caulimoviruses and the gemini viruses (section 4.1) command special interest since they are DNA viruses. While these have not been extensively developed as vectors, members of both groups have both had important roles in the development of techniques of genetic transformation.

The outer envelope of the cauliflower mosaic virus and the paucity of DNA sequences that can be removed without any detrimental effect have effectively limited its use as a vector. But molecular studies using this virus have yielded useful promoters, such as the strong, apparently constitutive, 35S promoter.

5. Transformation by direct gene transfer

Reports that cultured mammalian cells were capable of taking up and expressing cloned genes rapidly led to studies of the feasibility of introducing DNA directly to plant cells. The first successful example of direct gene transfer of defined DNA sequences to plant cells was the introduction of the large (120 kb) Ti plasmid from *A. tumefaciens* to petunia and tobacco protoplasts (Draper *et al.*, 1982; Krens *et al.*, 1982). Because the T-DNA conferred the capacity for hormone-independent growth on the cells, transformed tissue could be readily selected. As regenerated plants had modified hormone levels they tended to be developmentally abnormal and hence unsuitable for further studies of gene inheritance. However, this led the way to the development of smaller (5–10 kb) plasmids containing reliable selectable genes and reporter genes based upon *E. coli* plasmids. This avoided the introduction of genes encoding plant growth regulators and showed that Ti encoded genes, and in particular the 23 bp border sequences flanking the T-DNA region of the Ti plasmid, were not required for integration of foreign DNA into the plant genome.

5.1 Direct gene transfer to protoplasts

Isolated protoplasts were first used to demonstrate direct gene transfer to plant cells by determining transient expression of the introduced genes (Figure 2.2). Since then, methods developed for direct gene transfer to dicot protoplasts have been successfully applied to monocots and have led to the successful stable integration of the introduced DNA and regeneration of transformed plants (Horn *et al.*, 1988; Rhodes *et al.*, 1988; Zhang *et al.*, 1988). Transgenic plants of maize, rice, barley and orchard grass have now been obtained by this method.

The natural low permeability of the plasmalemma to macromolecules such as DNA must be increased for DNA to cross the membrane. This may be achieved either chemically or electrically. Two basic methodologies therefore exist for direct gene transfer to protoplasts. One method relies on chemical treatments of protoplasts with polyethylene glycol (PEG) at high pH in the presence of divalent cations either for delivery of naked DNA or DNA encapsulated in liposomes. The other method is based upon modification of membrane structure and permeability by application of high voltages, the so-called electroporation of protoplasts. Both of these methods have their own particular advantages and disadvantages, the

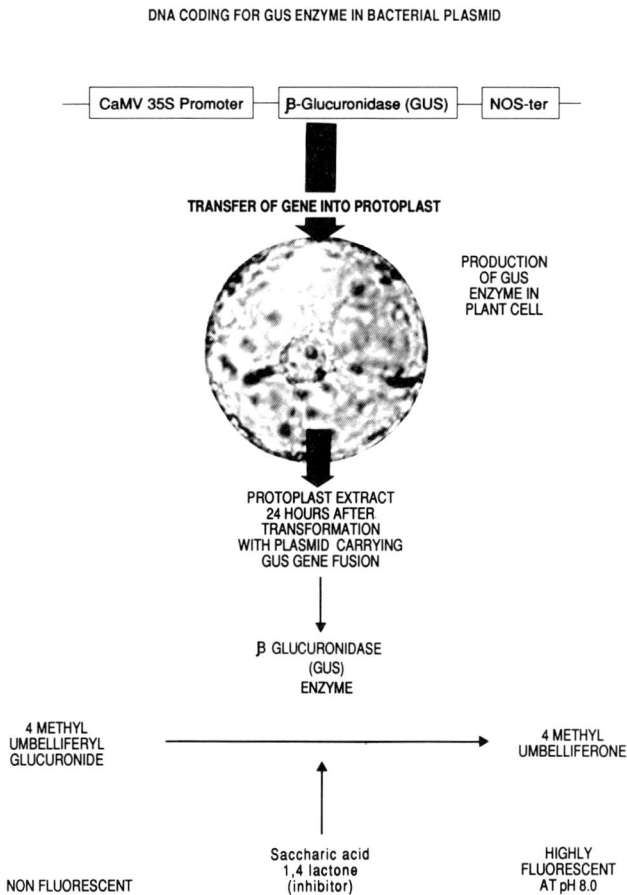

DNA CODING FOR GUS ENZYME IN BACTERIAL PLASMID

CaMV 35S Promoter — β-Glucuronidase (GUS) — NOS-ter

TRANSFER OF GENE INTO PROTOPLAST

PRODUCTION
OF GUS
ENZYME IN
PLANT CELL

PROTOPLAST EXTRACT
24 HOURS AFTER
TRANSFORMATION
WITH PLASMID CARRYING
GUS GENE FUSION

β GLUCURONIDASE
(GUS)
ENZYME

4 METHYL
UMBELLIFERYL
GLUCURONIDE

4 METHYL
UMBELLIFERONE

Saccharic acid
1,4 lactone
(inhibitor)

HIGHLY
FLUORESCENT
AT pH 8.0

NON FLUORESCENT

Figure 2.2 Principles of transient expression of β-glucuronidase gene in isolated protoplasts.

method of choice depending upon particular applications and species.

Transfer of the *nptII* gene by electroporation into protoplasts isolated from embryogenic cell suspension cultures of maize, followed by selection of kanamycin-resistant calli, resulted in stable transformation (at a frequency of 5% of viable protoplasts). Plants regenerated from these calli maintained and expressed the introduced gene but proved to be sterile (Rhodes *et al.*, 1988). The regeneration of transgenic rice plants has now been reported in several laboratories world-wide using both chemical (PEG) and electrical methods for DNA transfer to embryogenic protoplasts, followed either by antibiotic selection or by screening regenerated calli for *gus* gene expression without selection (Toriyama *et al.*, 1988; Zhang and Wu, 1988).

Chemical treatments

Polyethylene glycol at high pH has long been used to induce protoplast fusion and was used in pioneering studies to transfer Ti and *E. coli* plasmids to protoplasts (Krens *et al.*, 1982; Lörz *et al.*, 1985). The use of PEG to transfer DNA into the protoplasts has since been refined, with the result that rates of transformation are now many orders of magnitude higher than the original reported frequency of 10^{-5}. This has been achieved as a result of the following modifications.

1. Linearization of the plasmid DNA. In some systems, this has been shown to be 3 to 10 times more efficient for transformation, particularly for stable integration, than circular forms.
2. Heat shock treatment of the protoplasts. A 5-minute heat shock at 45 °C followed by transfer to ice immediately prior to addition of plasmid and PEG can increase efficiency of DNA transfer by several orders of magnitude. This has been particularly effective in DNA transfer via electroporation.
3. Optimization of divalent cations. Replacement of calcium phosphate and later calcium chloride in the protoplast transformation medium with magnesium ions immediately prior to the addition of PEG stimulates DNA uptake and transient gene expression (Negrutiu *et al.*, 1987).
4. Optimization of the PEG concentration and pH. Addition of PEG at concentrations up to 20%, at higher pH (up to pH 8), after, rather than before, addition of plasmid DNA has beneficial effects on transformation frequencies.
5. Addition of carrier DNA. Carrier DNA in the form of sheared calf thymus or salmon sperm DNA has been found to stimulate transformation frequencies in excess of that achieved with equal amounts of plasmid DNA. The mechanisms by which this promotes DNA uptake is not known and there is growing evidence that this carrier DNA may not be necessary if other parameters are optimal.
6. Manipulations following addition of PEG. Concurrent and gradual reduction in the PEG concentration and an increase in calcium ion concentration improves DNA transfer. Whether this is a result of improved protoplast viability or minimal exocytosis of DNA incorporated in cytoplasmic vesicles during PEG treatment is not known. The mechanism by which DNA crosses the plasmalemma membrane is discussed elsewhere (section 6.3).
7. Synchronization of division. There is some evidence that the stage of the cell cycle of recipient protoplasts can be important for achieving high transformation rates. Tobacco mesophyll protoplasts were synchronized by culture in aphidicolin (an inhibitor of DNA replication) and dichlorobenzonitrile (an inhibitor of cell wall synthesis), and were transformed with

the *nptII* gene. These protoplasts yielded up to 3% kanamycin-resistant colonies if they were transformed during the DNA replication (S) or mitotic (M) phases of the cell cycle. This was about twice the frequency at which unsynchronized protoplasts were successfully transformed (Meyer *et al.*, 1985). A 3–5-fold stimulation of transient expression levels in M phase tobacco protoplasts has also been reported.

DNA delivery via lipososmes

Fusion of negatively charged, DNA-containing liposomes (or spheroplasts of *Agrobacterium* or *E. coli*, carrying cloned DNA) with plant protoplasts was among the first methods used to achieve DNA transfer (Matthews and Cress, 1981; Hain *et al.*, 1984). Transformation frequencies of 4×10^{-5} were reported with no apparent rearrangement of the introduced DNA, although copies of the DNA were found to be in tandem. Reports of higher frequencies of transfer of naked DNA with PEG or electroporation have led to declining interest in this method of DNA delivery although an important aspect of this technique (which may be very useful for some systems) is the degree of protection afforded to the DNA by the liposomes, allowing efficient integration of intact plasmids of 9 kb or larger.

Electrical treatment (electroporation)

Application of high-voltage electrical pulses to protoplast suspensions increases the permeability of the plasmalemma to DNA. Above a critical field strength this leads to breakdown of the membrane, but below this a transient increase in membrane permeability can be induced. Because it has been suggested that this process results in transient pore formation in the membrane, the process has been termed electroporation. Electroporation therefore requires a balance between conditions that increase membrane permeability and conditions which result in membrane breakdown and loss of protoplast viability.

Electroporation has been used successfully for a number of years for protoplast transformation (Fromm *et al.*, 1986; Ou-Lee *et al.*, 1986) and has resulted in the development of two electroporation systems based either on low-voltage/long pulses or on high-voltage/short pulses. The situation is further complicated by the use of both square-wave pulse generators and the more commonly used capacitor discharge systems, which deliver exponentially decaying voltage pulses. Typical values for tobacco mesophyll are, for the low-voltage/long pulse method, 300–400 V cm^{-1} and 10–50 ms (exponential decay) and, for the high-voltage/short pulse method, 1000–15000 V cm^{-1} and 10 μs (square wave). Both methods have resulted in stably transformed cell lines and transgenic plants, with transformation frequencies as high as 2–8% in tobacco, particularly when high-voltage electroporation is combined with a low concentration (8%) of PEG during electroporation (Shillito *et al.*, 1985).

The optimal voltage and time constant depends on the protoplast system and the resistance of the medium but in general it is considered that, at least for short pulses, for every doubling of protoplast diameter, the voltage should be halved, although it should be noted that in general higher voltages may be needed for protoplasts isolated from suspension cultures than from mesophyll tissue (Shillito and Saul, 1988).

Transient expression of introduced genes has been found to depend not only on the physical parameters of electroporation, such as plasmid concentration, electrical conditions, protoplast size and density, but also on the physiological properties of the protoplasts. Low field strengths and long pulse durations are generally considered to give high rates of transient expression, while high field strengths and short pulses give higher rates of stable integration.

One of the most critical factors for increasing the permeability of the membrane and achieving high rates of DNA transfer is the size of the protoplast, or more particularly the size distribution of the protoplast population. The voltage required for membrane breakdown depends on protoplast size: the membranes of large protoplasts become permeable at low field strengths while those of small protoplasts respond to high field strengths. Thus, within a given population of protoplasts, which may be variable in size, only a small number of protoplasts will be transformed.

Preliminary experiments with new species or sources of protoplasts need to be carried out to determine survival (after 24 hour incubation) at different applied voltages. In general, conditions giving about 50% survival have been found to be most effective for stable transformation.

Optimized PEG-mediated transformation methods are now considered to be more reliable and efficient than electroporation for direct DNA transfer to protoplasts, although electroporation may still be the method of choice in systems sensitive to high PEG concentrations.

5.2   Direct gene transfer to cells and tissues

Recent developments in direct gene transfer have concentrated on delivering DNA into individual cells either in culture or in whole tissues. This obviates the need to remove the cell wall as in protoplast transformation methods and hence removes the associated problems of plant regeneration from protoplasts. Many of these techniques, however, have not been fully developed and only limited examples of their successful use are available. The general acceptance of these methods therefore awaits further research.

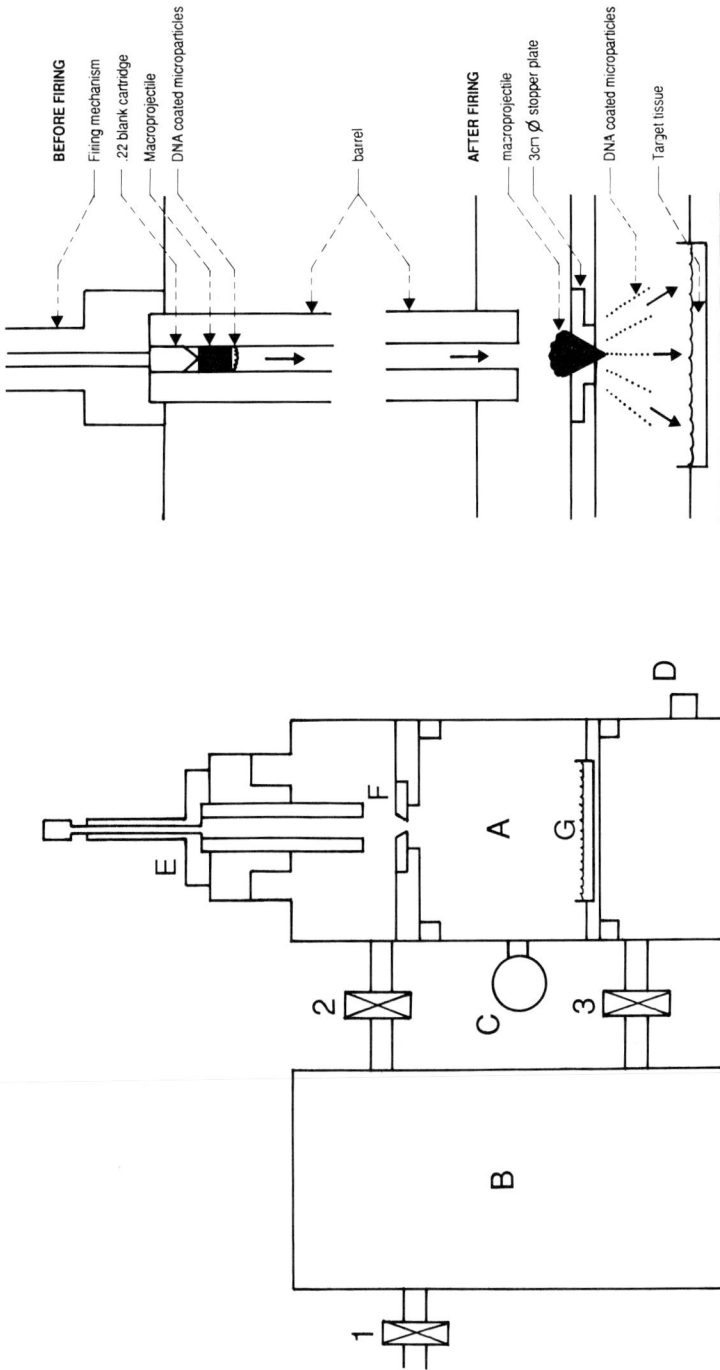

Figure 2.3. Construction and operation of DNA particle gun. A – sample chamber, B – ballast chamber, C – vacuum gauge, D – air inlet valve, E – firing mechanism, F – stopper plate, G – target tissue, 1, 2, 3 – vacuum taps.

Microprojectile bombardment (Biolistic^{TM} or DNA particle gun)

Microprojectile bombardment involves accelerating microprojectiles, typically tungsten or gold particles (1–2 μm in diameter), to velocities at which they can penetrate plant cell walls. This is achieved either electrostatically or with blank cartridges in a modified firing mechanism (Figure 2.3). Recent developments, however, utilize compressed air to accelerate the particles, giving greater control of particle velocities. DNA is bound to the particles and is therefore delivered into the cell. Firing is usually carried out under vacuum and precautions are taken to minimize cell damage by shock waves. Critical factors in increasing transformation frequencies include the number and velocity of the microprojectiles, the type of support used for the cells or tissues, and the concentration of spermidine and $CaCl_2$ used to bind the DNA to the particles.

This approach was first demonstrated using transient assays of gene expression in onion epidermis (Klein *et al.*, 1987) but later studies were extended to transient *gus* and *cat* expression in maize (Klein *et al.*, 1988, 1989), rice, soyabean and wheat (Wang *et al.*, 1988). Since then detailed studies have been carried out on a wide variety of species and tissues.

In combination with the *gus* reporter gene, which allows the visualization of transformed cells which are transiently expressing the gene, this method of direct DNA transfer provides a powerful tool for studying gene expression in a variety of cell types, including embryogenic cell clusters, leaf epidermis and mesophyll, and in cells of intact tissues such as apical meristems and somatic and zygotic embryos. The application of this technique has permitted rapid analysis of 5′ promoter regions for enhancer and silencer elements, the effect of introns and the role of 3′ terminator regions on gene expression and mRNA stability. When used for gene transfer to intact tissue, these techniques also allow rapid analysis of tissue-specific expression of gene constructs. Such analysis is needed to ensure normal regulation of gene expression prior to production of transgenic plants.

Recent evidence indicates that microprojectile bombardment gives similar results to DNA transfer via electroporation of protoplasts. As with other techniques for gene transfer to cell clusters or tissues, this method also suffers from the disadvantage that regenerated plants will almost certainly be chimeric for the introduced gene and hence genes introduced by this method may not be heritable.

Regulation of anthocyanin biosynthesis and pigmentation of maize aleurone provides an example in which microprojectile bombardment has been successfully used to study gene regulation in intact tissue. Anthocyanin biosynthesis is controlled by a number of structural and regulatory genes. The C and R alleles are regulatory genes required for expression of the Bronze 1 (*Bz1*) structural gene. Mutations in *Bz1* result in the production of a bronze colour in the aleurones. Transfer of a cloned *Bz1* gene by particle bombardment into cultured maize aleurones which have functional C and R genes but which are mutants for the *Bz1* gene successfully restored anthocyanin pigmentation in cells that expressed the introduced gene. Transfer of plasmids lacking the *Bz1* gene did not restore pigmentation. Furthermore, transfer of the *Bz1* gene into aleurones of mutants for C, R and *Bz1* did not result in restoration of pigmentation (Klein *et al.*, 1990). The construction of a chimeric gene consisting of the *Bz1* promoter region driving the firefly *lux* gene has also allowed regulatory sequences in the *Bzl* promoter to be studied independently of anthocyanin biosynthesis. These results demonstrate that microprojectile gene transfer is effective for studies of regulation of gene expression in intact tissues and that appropriate genetic control of the introduced genes occurs.

As a result of optimization of parameters determined from transient assays, transformation following microprojectile bombardment has been reported in soyabean, tobacco and maize at frequencies of 10^{-4} to 10^{-5} with 1 to 10% of transformation events resulting in stable integration. In the case of maize, embryogenic cell suspension cultures were bombarded as a thin layer on filter paper, with plasmid DNA carrying the *nptII* and *gus* genes. Selection of cells with kanamycin followed by screening of colonies for *gus* showed that intact cells of cereals can be stably transformed by microprojectile bombardment (Klein *et al.*, 1989).

The particle gun has also been used with pollen, early-stage embryoids, somatic embryos and meristems. Bombardment of soyabean meristems followed by plant regeneration has resulted in transgenic plants (McCabe *et al.*, 1988). Prospects also exist for utilizing this technique in direct gene transfer to cell organelles. Gene transfer to *Chlamydomonas* chloroplast and yeast mitochondria has been reported.

Microinjection of DNA into cells and protoplasts

Microinjection of DNA into cells using capillary micropipettes is the most direct and the most precise method of delivering DNA into specific cell compartments. Early attempts with plant cells followed successes reported with animal cell systems but were hampered by the cell wall or by the fragility of protoplasts. Methods were therefore developed to immobilize protoplasts in agarose or on glass with polylysine or by holding the protoplasts under suction. Successful transformation of wild-type Ti plasmid into alfalfa (Reich *et al.*, 1986) and tobacco protoplasts (Crossway *et al.*, 1986) was achieved initially but it rapidly became apparent that injection into the nucleus or the cytoplasm was necessary to achieve high rates of transformation. Stable transformation frequencies of 14–66% of injected protoplasts have been reported, but the absolute number of transformants is always low due to the low number of protoplasts which can be microinjected (maximally 40–50 protoplasts/hour).

Microinjection has now been successfully used with cells and proto-plasts of tobacco, alfalfa, oilseed rape and other species of brassica. The most successful example is the microinjection of plasmid DNA carrying the *nptII* gene into embryoids derived from microspores (12-cell stage) of rape (Neuhaus *et al.*, 1987). Of the 80% of embryoids regenerating to plants, half were stably transformed. Although many of these plants proved to be chimeric, they segregated *in vitro* through embryogenesis to give stable transformants. These and other results clearly demonstrate that cells in meristematic clumps capable of plant regeneration can be transformed by microinjection of DNA.

Microinjection into vacuolated, non-embryogenic cells and into protoplasts has not generally been successful, as much of the DNA is injected into the vacuole of these cells and is degraded. Microinjection of DNA into the nucleus is required for high levels of transformation. Microinjection of evacuolated protoplasts has been partially successful (Lawrence and Davies, 1985), both with plasmid DNA and with whole chromosomes (Griesbach, 1987). The latter application of microinjection provides a bridge between genetic transformation, using single known genes, and somatic hybridization, in which whole genomes are transferred. The major limitations to this method are that it is technically demanding, requires the availability of small clusters of embryonic cells and often results in chimeric plants.

Microinjection may also be of some value in transient assays where gene transfer to particular cell types can give information about cell-specific expression and cell lineage. A visual marker such as *gus* can be detected in a single cell.

Macroinjection of DNA into plants

The technique of injecting DNA directly into the inflorescences of plants was first applied to rye (De la Pena *et al.*, 1987) as a result of studies of the uptake of the low-molecular-weight compounds colchicine and caffeine by cells of rye inflorescences. This showed that, for a short time during floral development, generative cells were capable of taking up these compounds. Subsequent injection of DNA into developing tillers led to the apparent production of transformed plants at a low frequency (2 plants from 3000 seeds from 100 injected plants).

In this method, plasmid DNA containing a selectable antibiotic resistance gene is suspended in buffered medium and injected with a syringe needle directly into the lumen of the developing inflorescence. The resulting seeds are selected by exposing them to the antibiotic during germination. The hypothesis is that microspores at a specific time in their development are capable of taking up DNA.

Later work with other cereals, such as wheat and recently barley

(Mendel *et al.*, 1990), has been disappointing. Very low frequencies of transformation were reported in barley and no transformation occurred in wheat. In barley, *nptII* activity corresponding to about 1% of that of stably transformed tobacco plants was detected, but the genes were not stably inherited.

Electrical and laser modification of cell membrane permeability

Electroporation conditions which make protoplast membranes permeable also have potential for transfer of DNA into intact cells, although mild enzymatic treatment of the cells is usually necessary to allow DNA to cross the cell wall. To date only low levels of transient expression have been detected in sugarbeet (Lindsey and Jones, 1987) and maize cells (M.E. Fromm, personal communication) using this method. A further development, derived from electroporation of DNA into cells, is the electrophoresis of DNA into germinating seeds. Voltages of 2–10 V cm^{-1} at currents of 0.1 mA were applied for up to one hour to barley embryos (punctured by platinum electrodes) in contact with plasmid DNA containing the *gus* gene. *Gus* activity was detected following germination of the embryos (Ahokas, 1989). However, while transient expression was demonstrated histochemically, no molecular analysis for stable integration was reported.

The use of lasers to puncture holes in plant cell walls and membranes has also been demonstrated (Weber *et al.*, 1988), following its success in stably transforming animal cells at a frequency of 10^{-3}. While uptake of DNA into plant cells was demonstrated, no attempt was made to confirm transient expression or stable integration.

Transformation of pollen

There have been some reports that soaking pollen in DNA solutions prior to fertilization can result in gene transfer (Ohta, 1986). Such a simple and direct method of gene transfer would be of great value and would render current methods of gene transfer by *in vitro* manipulation redundant. However, no molecular confirmation was presented and later results, using cloned genes rather than total genomic DNA, which would be expected to give higher transformation rates, have not substantiated this method of gene transfer (Waldron, 1987). Currently, there is no direct evidence for the transfer of foreign DNA into pollen by soaking.

DNA transfer via the growing pollen tube

This technique was first developed using wheat and rice (Lu and Wu, 1988). Plasmid DNA carrying a selectable antibiotic resistance gene is simply applied to the cut surface of the stigma some time (5–20 minutes to 2–3 hours, depending upon the rate of pollen tube growth) after pollination. Resultant seeds are then screened for antibiotic resistance. In rice up to 20% of the seeds grew into plants which contained the *nptII* gene, with copy numbers varying from 1 to 300.

In similar work with barley, transformation frequencies of 10^{-3} to 10^{-4} in F1 seedlings were obtained, but only a low level of expression of the introduced *nptII* gene (driven by the *mas* or *nos* promoters) was detected and this was lost in mature plants and in the next generation (Mendel *et al.*, 1990).

DNA uptake into imbibing zygotic embryos

Recent experiments (Topfer *et al.*, 1989) have confirmed earlier reports (Ledoux *et al.*, 1974) of uptake and expression of plasmid DNA in isolated dry embryos during imbibition. Uptake of DNA by imbibing cereal (wheat, barley, rye) and legume (pea, bean) embryos was detected by monitoring the expression of an introduced *nptII* selectable marker gene. Amplification of this activity with a vector capable of autonomous replication (derived from the wheat dwarf virus) indicated that the observed expression was derived from uptake and expression in the plant embryo cells and was not due to contaminating micro-organisms.

It is difficult at the moment to compare the usefulness of this method for transient expression studies with other, better-characterized systems, such as those based on protoplasts or microinjectile bombardment, as more detailed studies are needed to determine the efficiency of transformation and the mechanisms of DNA uptake. However, as the embryos germinate at high frequency and give rise to plants, this method may also permit the stable transformation of plants.

The ability of DNA to pass through cell walls is somewhat unexpected. The 4 nm pore size of cell walls would be expected to be too small for passage of the 4.1- or 8.7-kb plasmid used. However, as the embryos are somewhat damaged during the isolation procedure (homogenization of seeds in cyclohexane/carbon tetrachoride), DNA may only need to cross plasmodesmata between intact and damaged cells. This, combined with the rapid flux of water and reassociation of membranes on hydration, may allow DNA to be taken up into the cells.

A recent modification of this technique involves the application of solution containing plasmids carrying the *nptII* gene to the exposed surface of an embryo, whose testa has been mechanically removed by grinding. This achieves both embryo rehydration and DNA transfer in one step. After a suitable incubation period seeds are then germinated on antibiotic selective media. This suggests that gene transfer to dry embryos can be achieved without the need to isolate zygotic embryos from seed.

Fibre-mediated DNA delivery to plant cells

Silicon carbide fibres (0.6 μm diameter × 10–80 μm long) coated with
DNA have recently been used to deliver DNA to cells of maize and
tobacco in suspension culture (Kaeppler *et al.*, 1990).

Cells from a suspension culture, when vortexed in the presence of
medium, plasmid DNA encoding the *gus* gene and silicon carbide fibres,
showed reproducible transient *gus* expression at an estimated frequency of
10^{-4} cells per treatment. Evidence that the fibres act as microinjection
needles puncturing the cell walls and delivering DNA into the cytoplasm or
nucleus was presented, but no further studies of rates of stable integration
or regeneration to plants were undertaken. This method, if found to be
widely applicable, must represent the most rapid and inexpensive method
yet devised for introducing DNA into plant cells.

5.3 DNA uptake into protoplasts and transfer to the nucleus

Opinion is divided about the mechanism by which DNA crosses the
plasmalemma and then enters the nucleus. However, under the influence of
the elevated osmotic pressures created by the use of high concentrations of
PEG during transformation of protoplasts, DNA may be incorporated into
membrane-bound vesicles in the cytoplasm in a similar manner to
formation of cytoplasmic vesicles on plasmolysis of cells. These may then
fuse either with the nuclear membrane systems continuous with the nuclear
membrane and effectively transfer DNA to the nucleus. Evidence for this
hypothesis comes from recent work on cation-mediated endocytosis and
exocytosis in oat protoplasts. These studies suggest that swelling and
shrinking of protoplasts during osmotic adjustment occurs by removal and
addition of membrane from the plasmalemma (Glaser and Donath, 1989).

DNA transfer by electroporation is thought to occur as a result of
transient changes in the lipid bilayer structure of the plasmalemma induced
by the electric field. The result is the formation of membrane pores through
which the DNA may enter the cell by electrophoresis. This hypothesis gains
support from the higher rates of DNA transfer that accompany the use of
capacitor discharge systems. These give rise to exponentially decaying
pulses thought to be conducive to electrophoresis of DNA through the
membrane pores created by the peak voltages. Electroporation using
square-wave pulses lacks this period of exponential voltage decay and,
although it may be ideal for creating conditions for protoplast fusion, it
may be less effective for DNA transfer.

6. Expression, integration and inheritance of inserted genes

6.1 Transient expression of genes introduced by direct gene transfer

The majority of the DNA introduced into cells by direct gene transfer is degraded and only a small fraction becomes integrated into the genome. However, this DNA can be expressed in the cell and forms the basis of transient assays. These assays are commonly used for the analysis of gene expression and for rapid evaluations of gene transfer efficiency, which are needed for the development of stable integration systems. Transient expression studies require an efficient method of delivering DNA into the cells and a simple assay for the gene product. For example, protoplasts treated with DNA under suitable uptake conditions are incubated for a time (usually 24–72 hours) to allow transcription and translation of the reporter gene (Figure 2.4). The activity of the gene product is then determined either chemically, by extraction of the protoplasts, or histochemically with suitable substrates which yield a visible product. Alternatively, the presence of the gene product can be determined using antibody techniques.

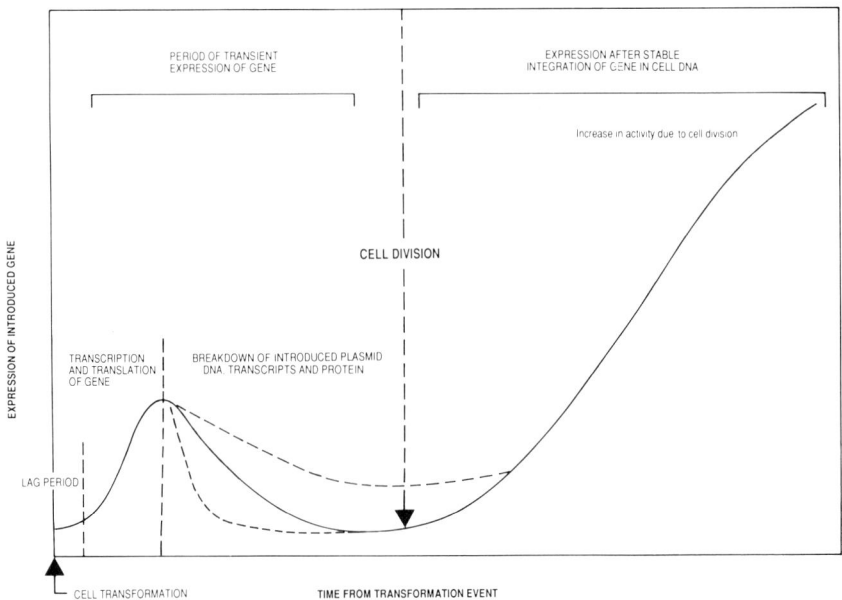

Figure 2.4. Relationship between transient expression and expression after stable integration of a gene introduced by direct gene transfer.

The analysis of gene expression by transient assays has been used to determine the efficiency of gene promoters and the role of different domains in the 5′ region, the role of introns and the effect of different 3′ regions. In addition, the expression of coding sequences and the stability of gene products can be rapidly determined. This is particularly useful for evaluation of selectable genes prior to use in studies of stable integration. Transient assays may also be used for the optimization of variables associated with different methods of direct gene transfer. Several reporter genes are now available (section 2.3) which are easy to assay, both in cell extracts and histochemically. Histochemical assays are useful for determining the frequency and type of cells which receive and express the introduced DNA.

6.2 Integration of genes introduced by direct gene transfer

One of the major disadvantages of direct gene transfer is the variable pattern of DNA integration which can occur. Vector DNA may be subjected to cleavage resulting in rearrangement, truncation, tandem formation and linkage of co-transformed genes either during or prior to integration. These modifications have frequently been observed to occur at higher rates than with *Agrobacterium*-mediated transformation.

Unequivocal integration of DNA into the nuclear genome, but not into the plastid or mitochondrial genomes, has been demonstrated using the majority of direct DNA transfer methods. Several copies of genes have been shown to occur at one site, but at different sites in different transformation events. Evidence for directed integration into a predetermined location in the plant genome by homologous recombination is also leading to the development of gene targeting systems in plants (Paszkowski *et al.*, 1988).

The initial copy number and arrangement of the introduced DNA in the genome may also be influenced by the method of DNA transfer. For DNA transfer to protoplasts, lowering the ratio of DNA to protoplasts has been found to decrease the transformation frequency per protoplast and to increase the frequency per amount of DNA. This also resulted in a decrease in the number of integrated copies from 5–10 at high DNA ratios to 1–5 at lower DNA ratios. Omission of carrier DNA, which in some systems stimulates transient expression and increases the number of integrated copies, has been shown to result in simpler integration patterns and lower numbers of copies integrated into tobacco protoplasts. Other methods of DNA delivery such as liposome-mediated gene transfer, microinjection and microprojectile bombardment have been shown to result in copy numbers varying from one to five.

The size of plasmid constructs which can be efficiently introduced by direct transfer is limited in some cases by the method of delivery. Several workers have used PEG-mediated delivery of Ti plasmids to protoplasts with demonstrable incorporation of the 20-kb T-DNA region, although most workers use plasmids of 5–15 kb. Liposomes have been used to deliver 9-kb plasmids, although packaging plasmids greater than 25 kb into liposomes may present some problems. Shearing of DNA may limit the size of plasmid that can be microinjected or transferred with the DNA gun, although plasmids of up to 16 kb have been transferred by this latter method. Thus, while 5–6 kb regions have been introduced into plant cells by several techniques without internal rearrangements, none of the DNA delivery methods yet developed have demonstrated delivery and expression of plasmids larger than 16–20 kb. The delivery of intact DNA sequences greater than 16–20 kb into the plant genome is therefore not yet possible with any reliability. However, this would be highly desirable, particularly for application of gene complementation techniques, in which DNA from genomic libraries could be used to complement and hence identify mutated genes. Recent reports of gene targeting in plants with directed integration into predetermined locations in the host plant genome and subsequent restoration of gene activity by homologous recombination (Paszkowski *et al.*, 1988) would suggest this could become a powerful technique for gene identification.

Co-transformation using direct DNA transfer methods has also been demonstrated on a number of occasions. *E. coli* plasmid-based vectors carrying different genes, when mixed prior to application of DNA transfer methods, have resulted in cell colonies containing both genes. Efficient co-transformation of both selectable (*nptII*) and non-selectable (*gus*) genes, both under the control of the CaMV *35S* promoter, has been demonstrated (Lyznik *et al.*, 1989), as has co-expression in regenerated plants.

6.3 Stable integration of introduced genes

Criteria initially used to demonstrate stable transformation included assays, such as resistance to antibiotics or herbicides, or phenotypic changes in the plant as a result of expression of particular genes. These, however, do not provide sufficient proof of transformation. The minimum proof of stable transformation must be the demonstration that the cell DNA contains an integrated copy (or a number of copies) of the introduced gene sequence and that this is stably maintained during cell division. In addition, heritability of the sequence in a Mendelian fashion should be demonstrated. Expression of the transferred gene via determination of appropriate mRNA, protein or enzyme activity provides additional verification. Thus, to confirm integration, Southern hybridization of genomic DNA to probes derived from regions of inserted sequence should be carried out. A

useful recent modification to this technique is the use of synthetic oligonucleotide primers, either for internal coding regions of the gene or for
promoter regions (if different promoters are used and need to be distinguished), and the use of the polymerase chain reaction to confirm the
presence of the introduced sequence or sequences in the genomic DNA.

A common problem, particularly with the use of *Agrobacterium*
species as vectors, is residual contamination of the tissue. Contamination
may lead to the false conclusion that the introduced DNA is integrated into
the plant DNA if expression of the gene is not determined. However, in
this case, screening DNA extracts with probes derived from outside the
T-DNA region of the Ti and Ri plasmid will confirm the presence or
absence of contaminating *Agrobacterium* species. When dealing with direct
DNA transfer to tissue in particular, care also needs to be taken to ensure
that transformation of microbial endophytes has not occurred and that
extra chromosomal plasmid replication is excluded.

6.4 Inheritance of introduced genes

Regeneration of fertile transgenic maize plants from cell cultures co-
transformed by bombardment with the *bar* and *gus* genes has been
achieved and shown to be reproducible and to result in stable inheritance
of the introduced genes (Gordon-Kamm *et al.*, 1990).

Progeny from A. tumefaciens-*transformed plants*

In *A. tumefaciens*-mediated transformation, the new genes are normally
incorporated into the nuclear genome, although exceptionally they may
insert into the chloroplast genome (De Block *et al.*, 1985). Within the
nucleus, DNA inserts at random, with the potential for many different
insertions occurring within the same nucleus (Zambryski *et al.*, 1989).
These insertions frequently occur in tandem arrays and can disrupt the
nuclear DNA.

Progeny from plants regenerated from both *A. tumefaciens* and *A.
rhizogenes* show that inserted genes are inherited as single dominant genes.
In tobacco, shoots regenerated from crowngalls have been successfully
grafted on to normal plants and found to be fertile. These grafts had lost
some if not all of the inserted DNA. A natural 'shooty' T-DNA mutant of
A. tumefaciens has permitted plant regeneration in tobacco (De Greve *et
al.*, 1982). The resulting plants showed the inheritance of the inserted
octopine gene, in a straightforward, Mendelian fashion.

Early work with disarmed *A. tumefaciens* vectors was also done mainly
with tobacco (Barton *et al.*, 1983; De Block *et al.*, 1984) and these plants
have been shown to transmit the foreign genes in a Mendelian fashion.

However, in subsequent generations, T-DNA can be transmitted to progeny but some instability has been observed, again in tobacco (Gheysen *et al.*, 1990). This was probably caused by methylation or rearrangements of an unstable T-DNA locus.

Segregation of Ri and binary T-DNA in progeny of A. rhizogenes *transformants*

Some *A. rhizogenes*-transformed plants show characteristic morphological aberrations such as wrinkled leaves, reduced flower size and an altered root system (Tepfer, 1984). These transformants frequently show reduced fertility (Tepfer, 1984; Visser *et al.*, 1989; Webb *et al.*, 1990). Analysis of the progeny has provided conflicting results. In tomato (Shahin *et al.*, 1986), tobacco (Sukhapinda *et al.*, 1987a, b) and oilseed rape (Boulter *et al.*, 1990), T-DNA originating from the Ri plasmid and the binary vector segregate away from each other during meiosis, while in potato (Visser *et al.*, 1989) the genes failed to separate. Thus successful production of genetically transformed progeny lacking the hairy root phenotype depends on the relative positions of random events. Unless the T-DNA can be directed to specific chromosomes, this will always pose a problem.

Conclusions

'Genetically engineered soybean, cotton, rice, oilseed rape, corn, sugar beet, and alfalfa are expected to enter the market by the year 2000' (Oxtoby and Hughes, 1990). Methodologies are developing more slowly in many other species because of a variety of factors, including lack of routine and reliable methods for regeneration of plants through tissue culture and the lack of universally applicable approaches to transformation. Another limiting factor is the availability of suitable single dominant genes.

Once the gene is inserted, the breeding work begins. The method of production of commercial varieties depends on the mode of propagation of the species: tubers are used in vegetatively propagated crops like potato, but in seed crops varieties are produced by either selfing plants to produce plants homozygous for the introduced genes, or, in a self-incompatible species, by a crossing programme involving crossing the transformed, heterozygous plant with normal, untransformed plants and then crossing selected, transformed progeny. In these ways, stable lines in which the new genes are expressed can be produced.

Prior to the release of any new varieties of genetically manipulated plants, many questions must be answered regarding any effect on the environment and the risks to consumers (both direct and indirect). All such releases are controlled by governments – in the UK by the Advisory Committee on Releases to the Environment (ACRE). Many field trials are underway, using species such as potato to assess the effects of growing these novel plants in the field and to assess any risk of gene transfer between genetically manipulated crops and their untransformed neighbours or close relatives. The main questions should address the risks associated with the newly inserted gene, and how it will behave in its new genetic background, rather than the method of its insertion.

Suggestions for further reading

Klee, H., Horsch, R. and Rogers, S. (1987) *Agrobacterium*-mediated plant transformation and its further applications to plant biology. *Annual Review of Plant Physiology* **38**, 467–486.

Old, R.W. and Primrose, S.B. (1989) *Principles of Gene Manipulation. An Introduction to Genetic Engineering.* Blackwell Scientific Publications, Oxford.

Weising, K., Schell, J. and Gunter, K. (1988) Foreign genes in plants: transfer, structure, expression and applications. *Annual Review of Genetics* **22**, 421–477.

References

Ahokas, H. (1989) Transfection of germinating barley seed electrophoretically with exogenous DNA. *Theoretical and Applied Genetics* **77**, 469–472.

An, G., Watson, B.D. and Chiang, C.C. (1986) Transformation of tobacco, tomato, potato and *Arabidopsis thaliana* using binary Ti vector system. *Plant Physiology* **81**, 301–305.

Barton, K.A., Binns, A.N., Matzke, A.J.M. and Chilton, M.-D. (1983) Regeneration of intact tobacco plants containing full length copies of genetically engineered T-DNA and transmission of T-DNA to R1 progeny. *Cell* **32**, 1033–1043.

Boulter, M.E., Croy, E., Simpson, P., Shields, R., Croy, R.R.D. and Shirsat, A.H. (1990) Transformation of *Brassica napus* L. (oilseed rape) using *Agrobacterium tumefaciens* and *Agrobacterium rhizogenes* – a comparison. *Plant Science* **70**, 91–99.

Callis, J., Fromm, M. and Walbot, V. (1987) Introns increase gene expression in cultured maize cells. *Genes and Development* **1**, 1183–1200.

Capone, I., Spano, L., Cardarelli, M., Bellincampi, D., Petit, A. and Constantino, P. (1989) Induction and growth properties of carrot roots with different complements of *Agrobacterium rhizogenes* T-DNA. *Plant Molecular Biology* **13**, 43–52.

Comai, L., Facciotti, D., Hiatt, W.R., Thompson, G., Rose, R.E. and Stalker, D.M. (1985) Expression in plants of a mutant *aro*A gene from *Salmonella*

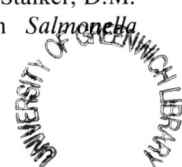

typhimurium confers tolerance to glyphosate. *Nature* **317**, 741–744.

Crossway, A., Oakes, J.V., Irvine, J.M., Ward, B., Knauf, V.C. and Shewmaker, C.K. (1986) Integration of foreign DNA following microinjection of tobacco mesophyll protoplasts. *Molecular and General Genetics* **202**, 179–185.

Dale, P.J., Marks, M.S., Brown, M.M., Woolston, C.J., Gunn, H.V., Mullineaux, P.M., Lewis, D.M., Kemp, J.M., Chen, D.F., Gilmour, D.M. and Flavell, R.B. (1989) Agroinfection of wheat: inoculation of *in vitro* grown seedlings and embryos. *Plant Science* **63**, 237–245.

De Block, M., Herrera-Estrella, L., Van Montagu, M., Schell, J. and Zambryski, P. (1984) Expression of foreign genes in regenerated plants and in their progeny. *EMBO Journal* **3**, 1681–1689.

De Block, M., Schell, J. and van Montagu, M. (1985) Chloroplast transformation by *Agrobacterium tumefaciens. EMBO Journal* **4**, 1367–1372.

De Block, M., Botterman, J., Vandewiele, M., Dockx, J., Thoen, C., Gosselé, J., Movva, N.R., Thompson, C., van Montagu, M. and Leemans, J. (1987) Engineering herbicide resistance in plants by expression of a detoxifying enzyme. *EMBO Journal* **6**, 2513–2518.

De Block, M., De Brouwer D. and Tenning, P. (1989) Transformation of *Brassica napus* and *Brassica oleracea* using *Agrobacterium tumefaciens* and the expression of the *bar* and *neo* genes in the transgenic plants. *Plant Physiology* **91**, 694–701.

De Greve, H., Leemans, J., Hernalsteens, J.-P., Thia-Toong, L., de Beuckeleer, M., Willmitzer, L., Otten, L., van Montagu, M. and Schell, J. (1982) Regeneration of normal and fertile plants that express octopine synthase, from tobacco crown galls after deletion of tumour-controlling functions. *Nature* **300**, 752–755.

Dekeyser, R., Claes, B., Marichal, M., Van Montagu, M. and Caplan, A. (1989) Evaluation of selectable markers for rice transformation. *Plant Physiology* **90**, 217–223.

De la Pena, A., Lörz H. and Schell, J. (1987) Transgenic rye plants obtained by injecting DNA into young floral tillers. *Nature* **325**, 274–276.

Dommisse, E.M., Leung, D.W.M., Shaw, M.L. and Conner, A.J. (1990) Onion is a monocotyledonous host for *Agrobacterium. Plant Science* **69**, 249–257.

Douglas, C.J., Halperin, W., Gordon, M. and Nester, E. (1985) Specific attachment of *Agrobacterium tumefaciens* to bamboo cells in suspension cultures. *Journal of Bacteriology* **161**, 764–766.

Draper, J., Davey, M.R., Freeman, J.R., Cocking, E.C. and Cox, B.J. (1982) Ti plasmid homologous sequences present in tissues from *Agrobacterium* plasmid-transformed *Petunia* protoplasts. *Plant Cell Physiology* **23**, 451–458.

Ellis, D., Roberts, D., Sutton, B., Lazaroff, W., Webb, D. and Flinn, B. (1989) Transformation of white spruce and other conifer species by *Agrobacterium tumefaciens. Plant Cell Reports* **8**, 16–20.

Evans, D.A. (1989) Somaclonal variation – genetic basis and breeding applications. *Trends in Genetics* **5**, 46–50.

Filatti, J.J., Kiser, J., Rose, R. and Comai, L. (1987) Efficient transfer of a glyphosate tolerance gene into tomato using a binary *Agrobacterium tumefaciens* vector. *Bio/Technology* **5**, 726–730.

Fromm, M.E., Taylor, L.P. and Walbot, V. (1986) Stable transformation of maize after gene transfer by electroporation. *Nature* **319**, 791–793.

Gheysen, G., Breyne, P., Müller, A., Depicker, A. and van Montagu, M. (1990) Stability of T-DNA inserts and their expression in plants. *Genetical Society Newsletter* (University of Birmingham, UK) **4**, 8–9.

Glaser, A. and Donath, E. (1989) Osmotically-induced surface area expansion in oat protoplasts depends on the transmembrane potential. *Journal of Experimental Botany* **40**, 1231–1235.

Gordon-Kamm, W.J., Spencer, T.M., Mangano, M.L., Adams, T.R., Daines, R.J., Start, W.G., O'Brien, J.V., Chambers, S.A., Adams, W.R., Willetts, N.G., Rice, T.B., Mackey, C.J., Krueger, R.W., Kausch, A.P. and Lemaux, P.G. (1990) Transformation of maize cells and regeneration of fertile transgenic plants. *Plant Cell* **2**, 603–618.

Graves, A.C.F. and Goldman, S.L. (1986) The transformation of *Zea mays* seedlings with *Agrobacterium tumefaciens*. Detection of T-DNA specific enzyme activities. Plant Molecular Biology **7**, 43–50.

Graves, A.E., Goldman, S.L., Banks, S.W. and Graves, A.C.F. (1988) Scanning electron microscope studies of *Agrobacterium tumefaciens*. Attachment to *Zea mays*, *Gladiolus* sp. and *Triticum aestivum*. *Journal of Bacteriology* **170**, 2395–2400.

Griesbach, R.J. (1987) Chromosome-mediated transformation via microinjection. *Plant Science* **50**, 69–77.

Grimsley, N., Hohn, T., Davies, J.W. and Hohn, B. (1987) *Agrobacterium*-mediated delivery of infectious maize streak virus into maize plants. *Nature* **325**, 177–179.

Hain, R., Steinbiss, H.H. and Schell, J. (1984) Fusion of *Agrobacterium* and *E. coli* spheroplasts with *Nicotiana tabacum* protoplasts – direct gene transfer from microorganism to higher plant. *Plant Cell Reports* **3**, 60–64.

Hauptmann, R.M., Vasil, V., Ozias-Akins, P., Tabaeizadeh, Z., Rogers, S.G., Fraley, R.T., Horsch, R.B. and Vasil, I.K. (1988) Evaluation of selectable markers for obtaining stable transformants in the Gramineae. *Plant Physiology* **86**, 602–606.

Hernalsteens, J.-P., van Vliet, F., De Beuckeleer, M., Depicker, A., Engler, G., Lemmers, M., Holsters, M., Van Montagu, M. and Schell, J. (1980) The *Agrobacterium tumefaciens* Ti plasmid as a host vector system for introducing foreign DNA in plant cells. *Nature* **287**, 654–656.

Hernalsteens, J.-P., Thia-Toong, L., Schell, J. and van Montagu, M. (1984) An *Agrobacterium*-transformed cell culture from the monocot *Asparagus officinalis*. *EMBO Journal* **3**, 3039–3042.

Herrera-Estrella, L., de Block, M., Messens, E., Hernalsteens, J.-P., van Montagu, M. and Schell, J. (1983) Chimaeric genes as dominant selectable markers in plant cells. *EMBO Journal* **2**, 987–995.

Hinchee, M.A.W., Connor-Ward, D.V., Newell, C.A., McDonnell, R.E., Sato, S.J., Gasser, C.S., Fischhoff, D.A., Re, D.B., Fraley, R.T. and Horsch, R.B. (1988) Production of transgenic soybean plants using *Agrobacterium*-mediated DNA transfer. *Bio/Technology* **6**, 915–922.

Hooykaas, P.J.J. and Schilperoort, R.A. (1984) The molecular genetics of crown gall tumorigenesis. In *Advances in Genetics 22. Molecular Genetics of Plants*, ed. J.G. Scandalios and E.W. Caspari, pp. 209–283.

Hooykaas-Van Slogteren, G.M.S., Hooykaas, P.J.J. and Schilperoort, R.A. (1984)

Expression of Ti plasmid genes in monocotyledonous plants infected with *Agrobacterium tumefaciens. Nature* **311**, 763–764.

Horn, M.E., Shillito, R.D., Conger, B.V. and Harms, C.T. (1988) Transgenic plants of orchard grass (*Dactylis glomerata* L.) from protoplasts. *Plant Cell Reports* **7**, 469–472.

Jordan, M.C. and McHughen, A. (1988) Glyphosate tolerant flax plants from *Agrobacterium* mediated gene transfer. *Plant Cell Reports* **7**, 281–284.

Kaeppler, H.F., Weining, G., Someus, D.A., Rines, H.W. and Cockborn, A.F. (1990) Silicon carbide fiber-mediated DNA delivery into plant cells. *Plant Cell Reports* **9**, 415–418.

Klee, H., Horsch, R. and Rogers, S. (1987) *Agrobacterium*-mediated plant transformation and its further applications to plant biology. *Annual Review of Plant Physiology* **38**, 467–486.

Klein, T.M., Wolf, E.D., Wu, R. and Sanford, J.C. (1987) High velocity microprojectiles for delivering nucleic acids into living cells. *Nature* **327**, 70–73.

Klein, T.M., Gradziel, T., Fromm, M.E. and Sanford, J.C. (1988) Factors influencing gene delivery into *Zea mays* cells by high velocity microprojectiles. *Bio/Technology* **6**, 559–563.

Klein, T.M., Kornstein, L., Sanford, J.C. and Fromm, M.E. (1989) Genetic transformation of maize cells by particle bombardment. *Plant Physiology* **91**, 440–444.

Klein, T.M., Goff, S.A., Roth, B.A. and Fromm, M.E. (1990) Applications of the particle gun in plant biology. In *Progress in Plant Cellular and Molecular Biology*, ed. H.J.J. Nijkamp *et al.* Kluwer Academic Publishers, Dordrecht, pp. 56–66.

Koukolíková-Nicola, Z., Shillito, R.D., Hohn, B., Wang, K., van Montagu, M. and Zambryski, P. (1985) Involvement of circular intermediates in the transfer of T-DNA from *Agrobacterium tumefaciens* to plant cells. *Nature* **313**, 191–196.

Krens, F.A., Molendijk, L., Wullems, G.J. and Schilperoort, R.A. (1982) *In vitro* transformation of plant protoplasts with Ti plasmid DNA. *Nature* **296**, 72–74.

Krieg, L.C., Walker, M.A., Senaratna, T. and McKersie, B.D. (1990) Growth, ammonia accumulation and glutamine synthetase activity in alfalfa (*Medicago sativa* L.) shoots and cell cultures treated with phosphinothricin. *Plant Cell Reports* **9**, 80–83.

Lal, R. and Lal, S. (1990) *Crop Improvement Utilizing Biotechnology.* CRC Press, Boca Raton, Florida, pp. 1–71.

Lawrence, W.A. and Davies, D.R. (1985) A method for the microinjection and culture of protoplasts at very low densities. *Plant Cell Reports* **4**, 33–35.

Ledoux, L., Huant, R. and Jacobs, M. (1974) DNA-mediated genetic correction of thiaminless *Arabidopsis thaliana. Nature* **249**, 17–21.

Lindsey, K. and Jones, M.G.K. (1987) Transient gene expression in electroporated protoplasts and intact cells of sugar beet. *Plant Molecular Biology* **10**, 43–52.

Lörz, H., Baker, B. and Schell, J. (1985) Gene transfer to cereal cells mediated by protoplast transformation. *Molecular and General Genetics* **199**, 178–182.

Lu, Z.-X. and Wu, R. (1988) A simple method for the transformation of rice via the pollen-tube pathway. *Plant Molecular Biology Reporter* **6**, 165–174.

Lyznik, L.A., Ryan, R.D., Richie, S.W. and Hodges, T.K. (1989) Stable co-transformation of maize protoplasts with gus A and neo genes. *Plant Molecular Biology* **13**, 151–161.

McCabe, D.E., Swain, W.F., Marinell, B.J. and Christou, P. (1988). Stable transformation of soybean (*Glycine max*) by particle acceleration. *Bio/Technology* **6**, 923–926.

Marks, M.S., Kemp, J.M., Woolston, C.J. and Dale, P.J. (1989) Agroinfection of wheat: a comparison of *Agrobacterium* strains. *Plant Science* **63**, 247–256.

Márton, L., Wullems, G.J., Molendijk, L. and Schilperoort, R.A. (1979) *In vitro* transformation of cultured cells from *Nicotiana tabacum* by *Agrobacterium tumefaciens*. *Nature* **277**, 129–131.

Matthews, B.F. and Cress, D.E. (1981) Liposome-mediated delivery of DNA to carrot protoplasts. *Planta* **153**, 90–94.

Mendel, R.R., Clauss, E., Hellmund, R., Schulze, J., Steinbiß, H.H. and Tewes, A. (1990) Gene transfer to barley. In *Progress in Plant Cellular and Molecular Biology*, ed. H.J.J. Nijkamp *et al.* Kluwer Academic Publishers, Dordrecht, pp. 73–78.

Meyer, P., Walgenbach, E., Bussmann, K., Hombrecher, G. and Saedler, H. (1985) Synchronised protoplasts are efficiently transformed by DNA. *Molecular and General Genetics* **201**, 513–518.

Negrutiu, I., Shillito, R., Potrykus, I., Biasini, G. and Sala, F. (1987) Hybrid genes in the analysis of transformation conditions. I Setting up a simple method for direct gene transfer in plant protoplasts. *Plant Molecular Biology* **8**, 363–373.

Nehra, N.S., Chibbar, R.N., Kartha, K.K., Datla, R.S.S., Crosby, W.L. and Stushnoff, C. (1990) Genetic transformation of strawberry by *Agrobacterium tumefaciens* using a leaf disc regeneration system. *Plant Cell Reports* **9**, 293–298.

Neuhaus, G., Spangenberg, G., Mittelsten-Scheid, O. and Schweiger, H.-G. (1987) Transgenic rapeseed plants obtained by the microinjection of DNA into microspore-derived embryoids. *Theoretical and Applied Genetics* **75**, 30–36.

Ohta, Y. (1986) High efficiency genetic transformation of maize by a mixture of pollen and exogeneous DNA. *Proceedings of the National Academy of Sciences (USA)* **83**, 715–719.

Old, R.W. and Primrose, S.B. (1989) *Principles of Gene Manipulation. An Introduction to Genetic Engineering.* Blackwell Scientific Publications, Oxford.

Ou-Lee, T.M., Turgeon, R. and Wu, R. (1986) Expression of a foreign gene linked to either a plant virus or a *Drosophila* promoter after electroporation of protoplasts of rice, wheat and sorghum. *Proceedings of the National Academy of Sciences (USA)* **83**, 6815–6819.

Oxtoby, E. and Hughes, M.A. (1990) Engineering herbicide tolerance into crops. *Tibtech* **8**, 61–65.

Paszkowski, J., Baur, M., Bogucki, A. and Potrykus, I. (1988) Gene targetting in plants. *EMBO Journal* **7**, 4021–4026.

Potrykus, I., Paszkowski, J., Saul, M.W., Petruska, J. and Shillito, R.D. (1985) Molecular and general genetics of a hybrid foreign gene introduced into tobacco by direct gene transfer. *Molecular and General Genetics* **199**, 169–177.

Puonti-Kaerlas, J., Stabel, P. and Eriksson, T. (1989) Transformation of pea (*Pisum sativum* L.) by *Agrobacterium tumefaciens*. *Plant Cell Reports* **8**, 321–324.

Rainer, D.H., Bottino, P., Gordon, M.P. and Nester, E.W. (1990) *Agrobacterium*-mediated transformation of rice (*Oryza sativa* L.) *Bio/Technology* **8**, 33–38.

Reich, T.J., Iyer, V.N. and Miki, B.L. (1986) Efficient transformation of alfalfa protoplasts by the intranuclear microinjection of Ti-plasmids. *Bio/Technology* **4**, 1001–1004.

Rhodes, C.A., Pierce, D.A., Mettler, I.J., Mascarenhas, D. and Detmer, J.J. (1988) Genetically transformed maize plants from protoplasts. *Science* **240**, 204–207.

Schäfer, W., Görz, A. and Kahl, G. (1987) T-DNA integration and expression in a monocot crop plant after induction of *Agrobacterium. Nature* **327**, 529–532.

Schmülling, T., Schell, J. and Spena, A. (1988) Single genes from *Agrobacterium rhizogenes* influence plant development. *EMBO Journal* **7**, 2621–2629.

Scowcroft, W.R. and Larkin, P.J. (1982) Somaclonal variation: a new option of plant improvement. In *Plant Improvement and Somatic Cell Genetics,* ed. Vasil, I.K., Scowcroft, W.R. and Frey, K.J. Academic Press, New York, pp. 159–178.

Shahin, E.A., Sukhapinda, K., Simpson, R.B. and Spivey, R. (1986) Transformation of cultivated tomato by a binary vector in *Agrobacterium rhizogenes*: transgenic plants with normal phenotypes harbor binary vector T-DNA but no Ri-plasmid T-DNA. *Theoretical and Applied Genetics* **72**, 770–777.

Shillito, R.D. and Saul, M.W. (1988) Protoplast isolation and transformation. In *Plant Molecular Biology: A Practical Approach,* ed. C.W. Shaw. IRL Press, Oxford, pp. 161–186.

Shillito, R.D., Saul, M.W., Paskowski, J., Müeller, M. and Potrykus, I. (1985) High efficiency direct gene transfer to plants. *Bio/Technology* **3**, 1099–1103.

Shimamoto, K., Terada, R., Izawa, T. and Fujimoto, H. (1989) Fertile transgenic rice plants regenerated from transformed protoplasts. *Nature* **338**, 274–276.

Smith, E.F. and Townsend, C.O. (1907) A plant-tumor of bacterial origin. *Science* **25**, 671–673.

Spena, A., Schmülling, T., Koncz, C. and Schell, J. (1987) Independent and synergistic activity of rol A, B and C logi in stimulating abnormal growth in plants. *EMBO Journal* **6**, 3891–3899.

Spencer, T.M., Gordon-Kamm, W.J., Daines, R.J., Start, W.G. and Lemaux, P.G. (1990) Bialaphos selection of stable transformants from maize cell culture. *Theoretical and Applied Genetics* **79**, 625–631.

Stafford, A. (1991) Genetics of cultured plant cells. In *Plant Cell and Tissue Culture,* ed. A. Stafford and G. Warren. Open University Press Biotechnology Series, Milton Keynes, pp. 25–47.

Sukhapinda, K., Spivey, R. and Shahin, E.A. (1987a) Ri plasmid as a helper for introducing vector DNA into alfalfa plants. *Plant Molecular Biology* **8**, 209–216.

Sukhapinda, K., Spivey, R., Simpson, R.B. and Shahin, E.A. (1987b) Transgenic tomato (*Lycopersicon esculentum* L.) transformed with a binary vector in *Agrobacterium rhizogenes*: non-chimeric origin of callus clone and low copy numbers of inegrated vector T-DNA. *Molecular and General Genetics* **206**, 491–497.

Tempé, J. and Casse-Delbart, F. (1989) Plant gene vectors and genetic transformation: *Agrobacterium* Ri plasmids. In *Cell Culture and Somatic Cell Genetics of Plants,* Vol. 6, Academic Press, London, pp. 25–49.

Tepfer, D. (1984) Transformation of several species of higher plants by *Agrobacterium rhizogenes*: sexual transmission of the transformed genotype and phenotype. *Cell* **37**, 959–967.

Topfer, R., Gronenborn, B., Schell, J. and Steinbiss, H.H. (1989) Uptake and transient expression of chimeric genes in seed-derived embryos. *Plant Cell* **1**, 133–139.

Toriyama, K., Arimoto, Y., Uchimiya, H. and Hinata, K. (1988) Transgenic rice plants after direct gene transfer to protoplasts. *Bio/Technology* **6**, 1072–1074.

Vasil, V., Clancy, M., Ferl, R.J., Vasil, I.K. and Hannah, L.C. (1989) Increased gene expression by the first intron of maize shrunken locus in grass species. *Plant Physiology* **91**, 1575–1579.

Visser, R.G.F., Hesseling-Meinders, A., Jacobsen, E., Nijdam, N., Witholt, B. and Feenstra, W.J. (1989) Expression and inheritance of inserted markers in binary vector carrying *Agrobacterium rhizogenes*-transformed potato (*Solanum tuberosum* L.). *Theoretical and Applied Genetics* **78**, 705–714.

Waldron, J.C. (1987) Pollen transformation. *Maize Genetics Newsletter* **61**, 36–37.

Wang, Y.-C., Klein, T.M., Fromm, M., Cao, J., Sanford, J.C. and Wu, R. (1988) Transient expression of foreign genes in rice, wheat and soybean cells following particle bombardment. *Plant Molecular Biology* **11**, 433–439.

Webb, K.J., Jones, S., Robbins, M.P. and Minchin, F.R. (1990) Characterization of transgenic root cultures of *Trifolium repens*, *Trifolium pratense* and *Lotus corniculatus* and transgenic plants of *Lotus corniculatus*. *Plant Science* **70**, 243–254.

Weber, G., Monajembashi, S., Greulich, K.O. and Wolfrum, J. (1988) Microperforation of plant tissue culture with a UV laser microbeam and injection of DNA into cells. *Naturwissenschaften* **75**, 35–36.

Weising, K., Schell, J. and Gunter, K. (1988) Foreign genes in plants: transfer, structure, expression and applications. *Annual Review of Genetics* **22**, 421–477.

White, D.W.R. and Greenwood, D. (1987) Transformation of the forage legume *Trifolium repens* L. using binary *Agrobacterium* vectors. *Plant Molecular Biology* **8**, 461–469.

White, F.F., Taylor, B.H., Huffmann, G.A., Gordon, M.P. and Nester, E.W. (1985) Molecular and genetic analysis of the transferred DNA regions of the root-inducing plasmid of *Agrobacterium rhizogenes*. *Journal of Bacteriology* **164**, 33–44.

Woolston, C.J., Barker, R., Gunn, H., Boulter, M.I. and Mullineaux, P.M. (1988) Agroinfection and nucleotide sequence of cloned wheat dwarf virus DNA. *Plant Molecular Biology* **11**, 35–44.

Zambryski, P., Tempé, J. and Schell, J. (1989) Transfer and function of T-DNA genes from *Agrobacterium* Ti and Ri plasmids in plants. *Cell* **56**, 193–201.

Zhang, H.M., Yang, H., Rech, E.L., Golds, T.J., Davis, A.S., Mulligan, B.J., Cocking, E.C. and Davey, M.R. (1988) Transgenic rice plants produced by electroporation-mediated plasmid uptake into protoplasts. *Plant Cell Reports* **7**, 379–384.

Zhang, W. and Wu, R. (1988) Efficient regeneration of transgenic plants from rice protoplasts and correctly regulated expression of the foreign gene in the plants. *Theoretical and Applied Genetics* **76**, 835–840.

Chapter 3
The Development of Genetically Modified Varieties of Agricultural Crops by the Seeds Industry

Richard J.A. Connett and Peter D. Barfoot
Agricultural Genetics Company Ltd, 154 Science Park, Milton Road, Cambridge CB4 4GG, UK

Introduction

During this century the yields of the major arable crops in Europe and the USA have approximately doubled, the result of a spectacular increase in agricultural efficiency. Plant breeders have made a major contribution to this increase through breeding and selection of new varieties. The main objectives of plant-breeding programmes are to increase yield potential, to increase pest/disease resistance and to improve quality. From their origins in basic research, the biotechnological methods of genetic diagnostics, tissue culture and genetic modification have been developed to a point where they now offer the prospect of enabling plant breeders to achieve further improvements in crop quality and yield potential. Although it is difficult to forecast when this new technology will have a major impact on the development of commercial products in a wide range of crops, a realistic assessment is that it will not be until the second half of this century or the start of the next. The extent to which these scientific and technical advances become translated into new varieties will be influenced by many factors; however, industry structure, technical progress, the scope for proprietary protection and the regulatory system will be particularly important. As these issues are complex and the legal frameworks for this technology are still developing, our aim is to give an overview of the main issues involved, particularly with respect to the registration of genetically modified agricultural crops in Europe and the USA.

1. Plant breeding and the seeds industry

The innovations from agricultural biotechnology will only be of direct use
to farmers when they appear in the form of seeds of new varieties. Seed is
delivered to the primary producer by the seeds industry; this includes
companies involved predominantly in plant breeding, those involved in the
retail sale of seed to the farmer, and those involved in both activities. The
organization and structure of this industry is highly complex. The industry
will both influence and be influenced by plant biotechnology. At this stage,
however, there are several characteristic features of the seeds industry in
the USA and Europe.

1. The market size is large in terms of both volume and value. Table 3.1
lists the world commercial seed markets for seven major crops in twelve
countries which in aggregate amount to more than US$5 billion. The
figure takes no account of the value of seed saved by farmers or of that
grown in the Third World. An important point here is that in terms of
commercial value in proportion to area planted, the hybrid crops pre-
dominate (compare maize and sunflower with other crops in Table 3.1).

2. Plant breeding is research intensive and product development timescales tend to be relatively long when compared to many other related
industries, for example the food industry. Development times approximate
to those in the pharmaceutical and agrochemical industries.

3. The industry is fragmented. Until quite recently the seeds industry was
characterized by a large number of relatively small companies. However, in

Table 3.1. The commercial seed markets for selected agricultural crops in the twelve
countries with the largest seed markets by value (1987). M = million.

Crop	Area planted (M ha)	Total seed planted (M tonnes)	Commercial seed purchased (M tonnes)	Value of commercial market ($M)	Seed value purchased/ unit area ($/ha)
Maize	58.3	1.1	1.0	2,495	47.1
Wheat	71.7	7.5	3.3	1,070	33.9
Soyabean	41.1	3.2	2.1	826	30.6
Barley	23.3	2.6	1.25	431	38.5
Sunflower	5.3	0.03	0.03	211	39.8
Oilseed rape	4.3	0.03	0.03	80	18.6
Cotton	7.0	0.02	0.01	65	18.6

common with many other industries, this is now changing rapidly. The last decade has seen many mergers and acquisitions, especially by large companies from the agrochemical and pharmaceutical sectors. Some observers predict that during the next decade the industry in the developed world will become dominated by a small group of companies operating globally (Kidd, 1985).

4. The industry is international. In some cases, although certainly not all, varieties have proved suitable for commercial use in a wide range of environments. Seed companies have exploited this by making arrangements to market their products as widely as possible, either themselves or more commonly through trading partners/agents. In addition, breeders often obtain sources of genetic variation (for example, resistance to diseases) from other countries and use countries in the opposite hemisphere to their own for field trials so as to provide two growing seasons per year.

5. The majority of the companies operating in this industry have relatively low levels of sustained profitability with notable exceptions such as those selling hybrid maize and sugarbeet. In general terms those companies able to control both plant breeding (and so receive royalties as the variety owner) and seed multiplication/distribution (giving a profit margin on seed sales) have been able to gain the maximum return.

In addition, national governments have historically, and for reasons of public policy, played a major role in the provision of new plant varieties. In the developed world this role is generally decreasing as a result of the encouragement of the private sector through the introduction of legal protection for new plant varieties. The USA, for example, now has a well-developed, private, plant-breeding industry as well as many state-controlled plant-breeding programmes funded by the Department of Agriculture. The involvement of governments in plant breeding has, however, influenced the expectations of the farmer in the price he/she is willing to pay for new varieties and may in turn affect the ability of the seeds industry to achieve a price premium for its biotechnologically derived products.

The seeds industry itself has been a major investor in plant biotechnology, but so have the agrochemical industry and technology transfer companies and, to a limited extent, the food-processing industry. The commercial rationale for investment by each of these is likely to be different. Agrochemical companies, for example, may be able to gain from increased sales of specific herbicides or maintain sales of agricultural inputs despite the pressure to reduce the use of agrochemicals. For crops such as tomatoes, sugarbeet, potatoes and vegetables, where there is an established linkage between food processors and the varieties which they require to be grown, the food industry stands to gain through reduced costs and improved quality. Increased profits to seed companies from the marketing of biotechnologically improved crop varieties may result either from a price premium or from an increase in market share. Companies offering their technology to seed companies stand to gain a share of such increased profits and for this reason biotechnology companies are often concerned to establish strategic relationships with food processors and seed companies. The possibilities for a price premium are based on the principle of added value, in which a proportion of the financial saving due to an incorporated trait is added to the price of seed of the variety. Although this principle is easy to apply for agrochemicals, where the potential loss in crop yield due to disease or pests provides the basis for how much the farmer will be prepared to pay for control, such calculations of added value are far more difficult to apply to seeds. This is the case for several reasons:

1. the complexity of varieties – varieties represent a complex balance of attributes;
2. the alternative varieties which may be available;
3. the ability of the farmer to save seed from commercial crops for replanting (the so-called 'farmer's privilege').

It is important to remember that a variety is not a product designed to be resistant to a single disease or to have a single quality characteristic. Every variety is an elaborate compromise between what is technically feasible for the breeder and the needs of the farmer, processor and consumer. Generally several varieties will meet the performance standards set by the variety testing authorities, although each variety will represent a slightly different balance of attributes. The farmer is thus able to choose between varieties and this forces seed companies to compete on the basis of price as well as on product characteristics. Also, in some crops the farmer is able to save seed for replanting. The extent to which this is done will depend on whether or not the crop is a hybrid (in which case the seed produced from sowing hybrid seed will not usually be suitable for replanting in the following generation because of genetic segregation), the possible contamination of the crop with weeds and other seeds which could cause a loss of yield or quality in the next generation, the risk associated with carrying seed-borne diseases to the next generation and, of course, price.

The extent to which seed companies will be able to obtain a financial premium for new biotechnologically improved varieties will thus be a function of: (i) the market demand for the particular attribute; (ii) the extent of vertical integration from specific variety to specific foodstuff or industrial product; (iii) the particular crop involved, particularly whether or not it is a hybrid; and (iv) the extent to which the attribute is proprietary rather than shared by competitor varieties.

2. Applications of biotechnology in plant breeding

Traditional plant breeding involves making large numbers of crosses between a range of parents which have been selected by the breeder for their desirable attributes. Some of the progeny from these crosses will show a combination of the best traits from the parents. After continued genetic recombination and a number of cycles of selection, a new variety is obtained which has a combination of desirable characteristics and which is distinct from any other variety. This process is simple in outline but complex in practice. Plant breeding involves a wide range of skills (Figure 3.1). Given this complexity it can take up to 15 years between the initial cross and the commercial release of a new variety. As a further complication there are various limitations to plant improvement through conventional breeding. Not only are plant-breeding programmes long-term

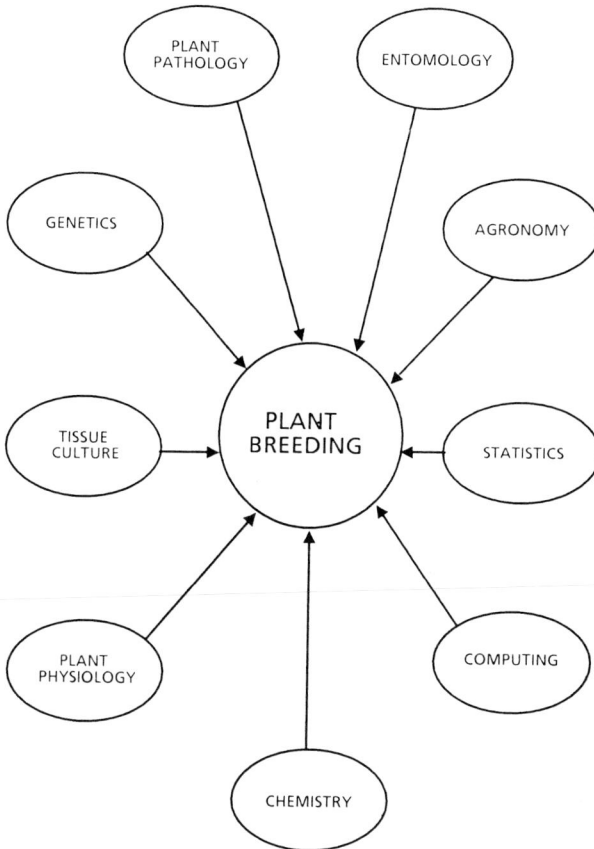

Figure 3.1. Technical disciplines involved in plant breeding.

and therefore expensive, they also require the cultivation and analysis of large numbers of plants which takes both time and space. Additionally, in some cases, the breeder has limited genetic variation for incorporation into new varieties. Thus techniques which overcome any of these constraints are of interest and potential value to plant breeders. Over the past decade technical advances in recombinant DNA and tissue culture methods have developed to a point where many analysts predict that plant biotechnology will have a major impact on the efficiency and success of conventional plant breeding (Gasser and Fraley, 1989). In the future the plant breeder may therefore need to broaden his knowledge base still further to include gene mapping, biochemistry and molecular biology.

In simple terms biotechnology can assist the plant breeder to improve crops in two ways:

1. by increasing the efficiency and effectiveness of selection;
2. by broadening the genetic base from which the breeder can select through direct or indirect genetic modification.

2.1 Techniques for increasing the efficiency and effectiveness of selection

Techniques for increasing the speed of selection include anther and pollen culture as a means, in certain crops, of producing true-breeding homozygous lines more rapidly than using techniques such as pedigree breeding or single-seed descent. The use of restriction fragment length polymorphisms (RFLPs) or other novel methods of genetic selection is also a potential way to increase the effectiveness of selection in the plant breeding process. RFLP analysis allows selection at the level of the genotype rather than selection at the level of the phenotype. This is likely to be an advantage for characters which are difficult to assess in the field such as resistance to certain diseases. Because these techniques can enhance the efficiency and effectiveness of selection, but do not directly lead to changes in the genetic composition of the plant, the final product (a new plant variety) will be indistinguishable from one produced using conventional methods. Thus no new regulatory issues are raised. As plant breeding has a long history and its products are generally recognized as safe (in the USA this phrase carries the acronym 'GRAS'), varieties developed through the use of such techniques will not generally need to satisfy regulations other than plant variety testing. It should be noted, however, that conventionally bred varieties of crops which are new to a particular country may require authorization for commercial production. Thus in the USA, technical information on oilseed rape ('canola') was examined by the Food and Drugs Administration before it was given GRAS status.

2.2   Techniques for broadening the genetic base for selection

Techniques for broadening the genetic base for the breeder to work with include genetic transformation and interspecific protoplast fusion. The products of such techniques are commonly referred to as genetically engineered organisms or genetically modified organisms (GMOs). By extension we refer to GMO crops and GMO varieties. Although the pace at which scientific advances have been converted into new products has been relatively slow, the rate at which new knowledge has been generated in this field has been dramatic (for a general overview see Lindsey and Jones, 1989). In particular, rapid progress has been made in isolating genes which confer characteristics such as herbicide resistance (see Chapter 4), insect resistance (see Chapters 6 and 7), virus resistance (see Chapter 8) and male sterility for hybrid seed production. Applied research on interspecific protoplast fusion has been particularly encouraging for transfer of male-sterile cytoplasms (for production of hybrid seed) in oilseed rape and rice, and may also be useful for transfer of disease resistances in crops such as potato and tomato. The extent to which such research will fall within the ambit of regulations for deliberate release of genetically modified organisms is still an open question and will be considered later. Research into genetic manipulation for crop improvement has been focused on the major field crops of the industrialized countries, including wheat, maize, rice, barley, sorghum, soyabean, alfalfa, cotton, sugarbeet, potato and tomato. These crops are all used to some extent in food manufacture. From this list of major crops, there are reports of successful transformation for all but wheat and barley. However, the transformation techniques which have been reported for the other crops still need to be improved so as to give high efficiency methods which are appropriate for a wide range of genotypes. For several of these crops, field tests of genetically modified crops which are potentially new varieties have been carried out already (Table 3.2).

Crops may also be developed for uses other than the nutrition of man and animals. Examples of this include plants producing oils for use in the manufacture of plastics and proteins for pharmaceutical use (so called 'molecular farming'). There are already reports of transformation of crop plants with genes for human serum albumin, pharmaceutical peptides and mammalian antibodies. This line of research may be particularly attractive for the biotechnologist because there may be significant commercial opportunities to dramatically reduce the production costs for therapeutic agents. As it is feasible that the market for certain pharmaceutical products could be satisfied by extraction from GMO plants produced in growth chambers (which are insulated from the external environment), the regulatory issues could be very different from those for GMO arable crops.

Despite the possibilities for using plant biotechnology to generate novel

Table 3.2. Targets for genetic improvement through biotechnology of major crops in which field test experiments have been carried out.

Crop	Trait
Alfalfa	Herbicide resistance Virus resistance
Cotton	Insect resistance Herbicide resistance
Oilseed rape	Herbicide resistance Hybrid production cytoplasmic male sterility nuclear male sterility Protein quality
Potato	Insect resistance Virus resistance
Soyabean	Herbicide resistance
Sugarbeet	Herbicide resistance Virus resistance
Tomato	Herbicide resistance Insect resistance Virus resistance Fruit ripening

plants for specialized uses, the greatest proportion of commercial investment in plant biotechnology research is focused on improvements of the major crops to satisfy existing markets. Many of the attributes which have been inserted into crops through genetic modification involve the improvement of agronomic characteristics which will be of primary value to the farmer. Field tests of genetic modifications which are designed to directly improve food quality are likely to follow shortly. It would thus seem that the applications of plant genetic modification are now at a threshold and that commercial organizations are starting to make detailed plans for wider field trials and marketing of GMO crops as products. The speed at which this threshold is crossed will depend critically upon the scope for protection of product innovations and the intensity of regulatory scrutiny. Commercial researchers working to develop GMO crop plant products will need to work through up to five tiers of official scrutiny and approval:

1. regulations on laboratory/glasshouse research;
2. applications for patent protection;
3. regulations on release into the environment of GMO crops (including

both small-scale research experiments and large-scale development trials);
4. variety protection and registration (in some cases the variety itself might be protected through patenting);
5. regulations on the food use of GMO crops.

The development of regulations for laboratory research will not be considered further, other than to suggest a general pointer to the development of regulations in this field. Over the last 10 years or so the regulations governing plant genetic manipulation in research laboratories have been made progressively less onerous as experience has grown and an excellent safety record has been maintained.

For steps 2, 3, 4 and 5, the main issues which must be addressed in seeking proprietary protection and regulatory approval for GMO varieties are extremely complex and vary considerably between countries. However, as 90% of the value of the annual retail seed consumption value in the developed free world is represented by just eleven countries (USA, Canada, Japan, Australia, France, Germany, Denmark, Holland, Italy, Spain and the UK) (Kidd, 1985), we have concentrated on the trading blocs which lead the development and the application of crop biotechnology: Western Europe and the USA. Although intellectual property protection and regulations are considered separately, at this stage they strongly influence each other. Thus if it takes 10 years to develop and obtain regulatory approval for a variety containing a patented gene, patent protection will only be available for approximately another 10 years (depending on the country). Similarly if the regulatory authorities require that results from field trials have to be disclosed publicly, then, in Europe at least, that information could not be used in a subsequent patent application.

3. Intellectual property protection for innovations in plant biotechnology

The developed countries have evolved legal systems which reward invention of products and processes by the granting of rights of limited monopoly for a defined period of time. These rights take a variety of forms and are generally called intellectual property rights (IPR). Patents, trademarks, trade secrets, copyright and plant variety rights are all types of IPR. For the seeds industry, the most important forms of IPR are plant variety rights (also known as plant breeders' rights) and patents. These two systems were designed to be mutually exclusive, with each having a particular type of subject matter. However, recent developments in biotechnology have blurred the distinction between them (Beier *et al.*, 1985; Bent *et al.*, 1987).

3.1 Plant variety rights

IPR protection raises particular difficulties in the field of plant breeding for two main reasons:

1. many of the technical advances are incremental, but nevertheless still worthy of protection. Such incremental changes are often not able to satisfy the standard criteria for patent protection;
2. there are difficulties in adequately describing the invention in such a way that it can be reproduced given the great complexity of biological systems.

For these reasons some countries have adopted a separate system of IPR for plant varieties – the plant variety rights (PVR) system. In essence this system protects innovations whereas the patent system protects inventions, which may or may not be translated into innovations. The international PVR system previously allowed the right to control production of seed of a registered variety for the purpose of sale (that is to say to legally block unauthorized production); recent developments have extended this right somewhat.

The PVR system is governed by the International Convention for the Protection of New Varieties of Plants (Union internationale pour la Protection des Obtentions Vegetales: UPOV) which was signed in 1961, amended in 1972, then revised in 1978 and in March 1991. This Convention binds 20 countries, including the USA, Japan and most countries of Western Europe. The membership comprises Australia, Belgium, Canada, Denmark, France, Hungary, Ireland, Israel, Italy, Japan, Netherlands, New Zealand, Poland, South Africa, Spain, Sweden, Switzerland, United Kingdom and the United States of America, although other countries have indicated that they intend to join. The most recent revision in the UPOV Convention has strengthened plant variety rights in the member countries, but individual member states are now required to incorporate its provisions into their national laws. The PVR monopoly agreed in the revised UPOV Convention extends to the production, processing and sale of a variety and may also be extended by national legislation to products derived directly from harvested material of that variety (e.g. oils from oilseeds and flour from wheat). The revised UPOV Convention also directs the member states to make available PVR protection in all plant species.

A potential variety for which PVR is sought must be examined officially in field trials and in some cases through chemical analysis in order to ensure that it meets the requirements of 'distinctness, uniformity and stability' (DUS). The DUS criteria ensure that the prospective new variety is different from other varieties and that it is both homogeneous and consistent in its essential attributes. In Europe such investigations are carried out by national authorities and involve trials over several years (two, three of four depending on the country and the crop) and in a range of trial sites. In Europe there is a secondary tier of testing of the value for cultivation and use (VCU). Only the best performing varieties of those which satisfy the DUS criteria will be approved for VCU. Once the variety has been approved by the regulatory authorities as satisfying the relevant criteria of DUS or DUS and VCU, seed of that variety is entered on a 'National List' and can be sold legally. The royalty rate applicable to multiplication of the variety for sale to the farmer will be set by the breeder. At present official trials in the European Community are conducted on a national basis, but draft legislation from the EC which was published in August 1990 is likely to establish a European PVR Office able to grant European PVR as an alternative to national PVR (O.J. No. C244/11-28.9.90).

The UPOV Convention contains what is known as a 'research exemption'. In essence this allows any plant breeder to freely use a registered plant variety for development of new plant varieties. There are restrictions, however, on direct use of varieties as parents in hybrid seed production and on making relatively small changes such that a second variety is 'essentially derived' from the first (as perhaps might be the case if a single gene is inserted into plants of a variety through genetic modification). The research exemption has undoubtedly contributed to the rapid development of new plant varieties but, as explained later, some argue that it is not equitable for this right to extend to plant varieties which involve inventive (and thus potentially patentable) processes. Another important provision is that national governments reserve the right to allow farmers to retain seeds of a crop for replanting, the so-called 'farmer's privilege'. The protection available is therefore less extensive than is provided by the patent system.

In contrast to the process of producing a new plant variety through conventional plant breeding, some of the products from biotechnology research (for example, a gene conferring a trait of major agronomic importance) could be used in the production of a wide range of plants in different varieties, species or families. This means that such new plants are the result of an inventive step which has wider potential application than conventional plant breeding which is limited to the crossing of sexually-compatible plants. The inventors of such advances in genetic modification have often sought to protect their work through the patent system. This will be considered next.

3.2 Patent protection

Many recent advances in plant biotechnology involve the application of microbiological techniques in genetic modification (the significance of 'microbiological' will become clear later). Some of these are the subject of patent applications. As there are considerable backlogs of patent applications under examination in most countries, relatively few patents have as yet been granted; this situation is, however, likely to change rapidly.

A patent is a form of property right granted by a state authority in respect of an invention and legally enforceable by its owner against unauthorized exploitation by others. The philosophy behind this system is that the patentee is granted a monopoly on his invention for a limited period of time in return for which the invention is disclosed to the public. At the end of the monopoly period, the patent may be worked freely by the public. In order to obtain patent protection, several criteria must be met: the invention must be new, useful and non-obvious. In addition the invention must not fall within any of the categories of invention which are specifically excluded from the grant of a patent. The invention must also be described sufficiently in the specification so as to enable it to be re-created by a person 'ordinarily skilled in the art'. In some cases it is necessary to deposit organisms used in the invention so that the invention can be recreated by this hypothetical person. It should be noted that there are important differences in the patent laws of the major trading blocks. For example, in Europe the essential principle is that any patent protection will be awarded to the 'first to file' whereas in the USA the date of invention rather than the date of first filing is of great importance. The exclusive rights of a patentee encompass production according to the invention, sale of patented products and sale of products which result directly from the use of a patented process. The protection given by a patent is determined by the scope of the 'claims' which are legal definitions of the technical product or process invented.

Given the complexity and rapid technical advances in plant biotechnology, there are strong reasons to suppose that the criteria for obtaining patent protection can be met. The types of inventions in this field for which patent applications have been filed include: tissue culture and micropropagation methods, methods for protoplast fusion, techniques for gene insertion, vectors, isolated genes and gene promoter and terminator sequences. The claims usually extend to plants developed using such methods or containing such genes or parts of genes. Thus several patent applications have claims of the type:

> Plants of species A having resistance to insect
> pest B through genetic transfer of gene C.

Although this claim relates to a plant, not a plant variety, the legal position of ownership of IPR in a variety derived from a patented plant is not yet entirely clear.

3.3 Potential conflict between PVR and patent protection

The PVR and patent systems for legal protection of genetically modified plants are at least potentially conflicting in several respects. On the basis of the UPOV Convention the results of biotechnology would at first sight be available for use by others once a variety in which these results have been incorporated has been marketed. In a relatively short time, it would be possible for other breeders to develop varieties having the same characteristic, but without paying anything to the inventor. This does not seem to be an equitable situation; doing the same thing in other technical fields would normally be considered an infringement of IPR.

On the other hand, there could be situations where an established plant variety is transformed with a gene so as to make a new variety which satisfied the DUS criteria. As an example of this, insertion of virus resistance genes into conventionally bred potato varieties might be a sufficient technical step to justify a claim for variety rights on the genetically modified variety. Not surprisingly, conventional plant breeders have been concerned that their varieties could be pirated by the biotechnologists without any recompense. In fact this possibility has been allowed for in the latest revision of the UPOV Convention. This has clearly defined provisions for 'dependency' which require those breeders who make relatively small changes to a pre-existing variety which nevertheless satisfy the DUS criteria to obtain permission from the owner of the original PVR before marketing the new variety. The original breeder has the right to refuse permission. The expectation, however, is that if the modification has commercial value, the parties will agree to share the royalty payments in a way which reflects their respective contributions.

Although both the USA and the EC are signatories to the UPOV Convention, the legal bases of IPR for varieties and biotechnological inventions have some important differences. This will be considered next.

Situation in the European Community

Most of the countries in Western Europe are signatories of the European Patent Convention (EPC) which came into force in 1978. The EPC does not have exactly the same membership as the EC; Ireland is not yet a member of the EPC whereas several non-EC countries are members. The EC has sought to rationalize this situation through a separate Community Patent Convention; this has not yet been brought into force.

The EPC treaty refers to the patenting of plant varieties in the following way:

> *Article 53 – Exceptions to patentability*
> European patents shall not be granted in
> respect of:

(b) plant or animal varieties or essentially biological processes for the production of plants or animals; this provision does not apply to microbiological processes or the products thereof.

However, some member countries such as Germany make exceptions to their national laws so as to allow patenting of varieties which are not protected by plant variety laws. In addition, as Article 53(b) says quite clearly that the exclusion does not apply to microbiological processes or the products of such processes, and it can be argued that all genetic engineering processes for producing new plants involve an essentially microbiological process, new genetically engineered plants should be patentable. Thus it seems that under the EPC and the corresponding national laws of member countries, patent claims will be granted on processes for and products of the biotechnological improvement of plants.

Recent case law has shown that the European Patent Office (EPO) intends to interpret the description 'variety' in a narrow sense, thus allowing patents to be granted on novel plants. More importantly, the EPO recently (March 1989) announced that it would grant patent for a plant gene expression method (Hall *et al.*, 1984). As well as protecting the method of manipulating the specific gene, the claims also apply to the plants themselves. The broadest product claim (Claim 1) in the patent is directed to:

> A DNA shuttle vector comprising T-DNA having inserted therein a plant gene comprising a plant promoter and a plant structural gene, the plant promoter being adjacent to the 5′ end of the plant structural gene and the plant structural gene being downstream from the plant promoter in the direction of transcription.

Subsequent claims relate to methods for using the shuttle vector and Claim 19 is directed to:

> A plant, a plant tissue, or a plant cell produced according to the method in any of the Claims 10–18.

The EPO has decided that this claim is allowable and does not fall within the exclusions of Article 53(b) of the EPC. This still leaves the question, however, of whether or not the EPO was correct in its initial judgement. One way in which this rather confusing situation can be clarified is through appeals to the patent authorities and ultimately litigation in the courts. A diverse range of organizations, from multinational food and chemical companies through to political and environmentalist groups, have lodged their opposition to the granting of this patent by the EPO. For some the argument is that the European Patent Convention was not intended to apply to higher organisms whereas for others the argument is that the broadest claim cannot be justified from the specific experiments and data provided in the specification.

The language of patent claims can also give rise to difficulties in interpretation. For example, in the type of claim illustrated in Claim 1 of the plant gene expression patent, does the phrase 'a plant promoter' mean a promoter which is active in a plant or one which is derived from a plant? If the latter is true then a promoter derived from a virus might be outside the scope of the patent. Although such issues might appear to be merely semantic, they could have major implications for the ability of inventors in biotechnology to extract a commercial return from their work. In future we may see litigation on such points as has happened in the pharmaceutical sector.

Situation in the USA

Under US patent law, plant patents can be granted for varieties which are reproduced asexually. Under the Plant Variety Protection Act special titles of protection have been available since 1970 for sexually reproduced varieties. Until relatively recently, however, the Patent and Trademark Office (PTO) denied utility patent protection for plants which reproduce through seed. Their argument, in simple terms, was that Congress would not have passed the Plant Variety Protection Act if the Patent Act had been intended to apply to seeds. That narrow interpretation was over-turned in September 1985. The US Board of Patent Appeals and Inter-ferences (ex parte Hibberd) quoted the Supreme Court decision in a previous case which concerned a novel organism in which it was declared that 'anything under the sun which is man-made' is potentially patentable. The patent application contained claims for maize tissue cultures, maize seeds and maize plants. The PTO Examiner had rejected these claims but the Board over-turned this argument. The decision removed the legal bar to patenting of varieties of sexually-propagated plants so that both PVR and patent protection is now available for varieties in the US.

3.4 Resolution of the potential legal conflict between PVR and patent protection

It has been suggested that the potential conflict between PVR and patent protection should be resolved by new legislation specifying that patent rights on a gene (which would include a plant containing a patented gene) would be exhausted once the patented gene had appeared in a registered variety. Another proposal is that patent holders should be forced to grant licences to plant breeders who have used the patented process or product from a variety in the production of new varieties – the plant breeder would be entitled to a 'licence of right'. As they would severely restrict the ability of patent holders to maintain exclusivity for their inventions, neither of these proposals seem to be fair to the patent holder.

In Europe this issue is likely to be resolved through an EC directive on the legal protection of biotechnological inventions which was proposed originally in 1988 (O.J. No. C10/3-13.1.89) but has not yet been enacted. This directive is intended to develop and harmonize the laws of the member states with respect to this and other issues of patenting in biotechnology. The directive as proposed confirms the availability of patent protection for plants, plant parts and plant materials, but excludes plant varieties from patent protection. Furthermore the draft directive provides for a royalty bearing 'dependency licence' to the available to the breeder of a new variety which incorporates patented features derived by breeding from a registered variety. In the opinion of many these provisions should be deleted from the draft so that the patent owner has the right to block unauthorized exploitation of his patented technology by plant breeders. This would mirror the dependency provisions in the revised UPOV Convention. Again the expectation is that, if the patented invention has value, the parties would negotiate licences with royalties shared in relation to their respective contributions.

4. Regulation of the release of genetically modified crops into the environment

4.1 Safety assessment

The rules governing the assessment of environmental safety of GMO crops are still evolving. Although it can be argued that product regulation should be based on the phenotype of the plant rather than on its method of genetic construction, national and international authorities have generally taken the view that the release into the environment of products derived from certain techniques should be subject to specific regulation. This is largely because of the pragmatic need to define the products and processes which are to be regulated so that the law is clear and consistent. This, of course, leads immediately to the question of which techniques should be defined as involving genetic modification. As an example of this, the authorities in Denmark on the one hand, and France on the other, seem to have different attitudes to whether or not the products of interspecific protoplast fusion should be classified as GMOs.

Although the history of plant breeding, that is to say essentially 'random' genetic modification, is such that examples of adverse environmental impact are extremely rare, it is generally accepted that the environmental safety of 'directed' genetic modification should be evaluated cautiously and thoroughly. Concerns revolve primarily around the ability of GMOs to replicate, from which it can be argued that any adverse effects would be multiplied and become uncontrollable. For this reason the authorities have adopted a cautious, case-by-case and step-by-step approach. Releases of GMOs have so far been on an experimental scale only but have produced no effects known to be detrimental to the balance of the environment.

The specific concerns of the scientific and regulatory communities in terms of effects on the environment can be summarized as three questions.

1. Will the process of genetic modification make the plant more persistent in the environment or more invasive of natural habitats?
2. Will the genetically modified plant pass on the inserted gene by pollen transfer or any other means in such a way that other plants become more persistent or invasive?
3. Will the genetically modified plant, or any plants to which the relevant genes have been transferred, have an additional character which could lead to detrimental effects in the environment?

These issues have provoked considerable debate (Boyce Thompson Institute for Plant Research, 1987; Council of the National Academy of Sciences, 1987; National Research Council, 1989; Tiedje *et al.*, 1989; US Congress Office of Technology Assessment, 1988). For example, a workshop held at the Boyce Thompson Institute for Plant Research in 1987 considered the potential risks from engineering the traits of herbicide resistance, disease resistance, insect resistance and altered storage product composition into the major North American crops (Boyce Thompson Institute for Plant Research, 1987). The conclusion from this meeting was that in most cases the products of biotechnology pose no risks significantly different from those which are well established from conventional plant breeding. It was also concluded that the possibility of major North American crops becoming weeds through genetic engineering was remote.

Nevertheless, appropriate procedures are needed to address any concerns and to identify, assess and minimize the potential risks. Both national governments and international organizations have sought to develop such regulatory procedures, although deciding how best to regulate the environmental release of GMOs has proved to be a controversial issue. Eminent organizations such as the US National Research Council (1989) and in the UK the Royal Commission on Environmental Pollution (1990) have reviewed these issues and suggested specific points which should be considered in establishing guidelines, regulations and laws in this field. An example of a simplified flow chart for evaluating any environmental effects that might result from releasing GMO plants is shown in Figure 3.2.

4.2 International aspects of regulation

Experiments involving the release of GMOs into the environment have been carried out in a number of countries including Australia, Belgium, Canada, Denmark, France, Germany, the Netherlands, Italy, the UK and the USA. Each of these countries has developed its own procedures and policies for regulating GMO field trials, usually with existing legislation as the basis. It is recognized that piecemeal international regulation is against the common economic interest and, although it has no formal regulatory role, the Organization for Economic Cooperation and Development (OECD) has sought to encourage common regulatory criteria and procedures which can be adopted by its member states. In its first report on this issue (OECD, 1986), the OECD considered that the development of general international guidelines governing the agricultural and environmental applications of GMO organisms was premature. The OECD did specifically recommend, however, that GMO applications should be assessed on a case-by-case basis by an expert group of scientists and that the development of GMO products should be carried out in a stepwise manner moving from the laboratory to the growth chamber, to the greenhouse, to limited field testing, to large-scale field testing. These recommendations have largely been followed with the result that technical knowledge of how safety assessments can be carried out has increased. The OECD has also recently published a discussion paper on 'Good Developmental Practice' for small scale field trials with GMOs.

4.3 Regulation in the European Community

In April 1987 the European Community published proposals for Community-wide regulation of the deliberate release into the environment of GMOs. The main feature of the proposal was that each release application should contain an analysis and assessment of possible risks and that the release should only be carried out following the approval of a competent national authority. The proposed directive was the subject of considerable debate within the European Commission and Parliament with opposition from some quarters on the grounds that its provisions would impede technical progress and from others that the regulations are not sufficiently stringent. At one point a proposal for a five-year moratorium on releases into the environment of GMOs was defeated in the European

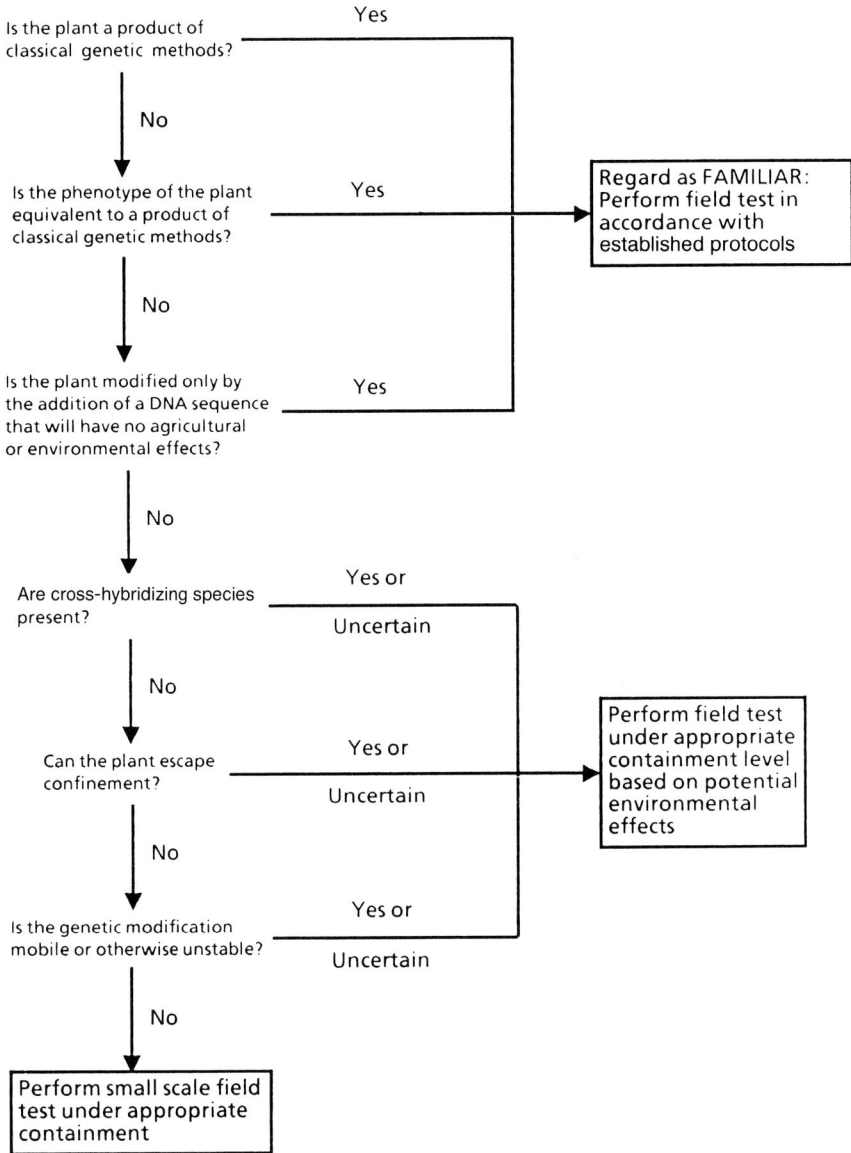

Figure 3.2. A framework for field testing of genetically modified plants. (Adapted from National Research Council, 1989).

Parliament by only two votes. Nevertheless, in May 1990, the member states approved this directive (O.J. No. L117/15-8.5.90), and they now have 18 months in which to translate its provisions into national laws. Essentially the directive provides that both experimental and commercial product releases of GMOs will be reviewed by competent national authorities. Applications for marketing of GMO products will be considered by the national authorities of all of the member states. In the event that one or more national authorities object to the marketing of a particular GMO product, the European Commission will decide whether or not the product should be approved throughout the whole of the EC. Fortunately the science has not had to wait for the legal systems to become established and technical progress has continued even in countries with a particularly sensitive climate of opinion on biotechnology; field trials of genetically modified petunia in Germany and herbicide-resistant sugarbeet in Denmark were carried out in 1990.

4.4 Regulation in the USA

In the USA assessment of the environmental safety of genetically modified plants falls within the jurisdiction of the United States Department of Agriculture's (USDA) Animal and Plant Health Inspection Service (APHIS) and the Environmental Protection Agency (EPA). In an effort to dovetail the activities of these regulatory bodies, the Office of Science and Technology published in 1986 a 'coordinated framework for the regulation of biotechnology' (Office of Science and Technology Policy, 1986) which called on the federal agencies to prepare guidelines for GMO field trials. This task has proved to be extremely contentious. The major issue to be decided is how to define the organisms and manipulations which should fall within and outside the regulations for organisms with 'deliberately modified hereditary traits' (Office of Science and Technology Policy, 1990; Department of Agriculture, 1991). One proposal, for example, is to exempt various categories including plants that result from natural reproduction or traditional breeding techniques, vascular plants regenerated from tissue culture, organisms modified by non-coding sequences that cause no phenotypic changes, organisms resulting from rearrangements within a single genome and organisms with new phenotypic traits conferring no greater risk to the environment than the safe parental strain. In some cases, authorities in individual states have also adopted a regulatory role and have passed specific legislation to control biotechnology. In addition, the US Congress may pass new legislation to govern the regulatory activities of the federal agencies. The regulatory system thus appears extremely complex (see Purchase and Mackenzie (1990) for a summary) and is still developing. However, despite this apparent complexity, many field trails have in fact been performed successfully and the number of approvals for such experiments is increasing rapidly.

4.5 Further use of GMO plants in plant breeding

After gaining permission to grow GMO plants in the field in order to evaluate their agronomic performance, the breeder will need to decide whether to register the GMO plant as a variety or to use the plant as parental material in a crossing programme. Given the steady increase in yield from conventional plant breeding, GMO varieties which are held up for several years in gaining regulatory approval for environmental release could experience difficulties in reaching the standards necessary for approval by the variety registration authorities. Thus, from the plant breeders' point of view, it would seem appropriate that once a GMO plant with a particular gene or genes which have been added has been shown not to have any detrimental effects on the environment, the results of further experimentation from multi-site trials and crosses with other plants should no longer be subject to specific environmental release regulations (other than a 'duty of care' which obliges the breeder to take all reasonable care to ensure that field experiments are safe). If this does prove to be the case, the breeder will be able to combine the particular trait conferred by the added gene with the other qualities of his best available germplasm in a potential new variety. The breeder will then have to submit the variety for DUS testing as outlined previously.

A related issue which must also be resolved is whether or not variety registration will involve safety assessment of the use of the plant product in foods. Whereas the legal and procedural aspects of environment impact assessments for GMO crops are now reasonably clear, the detailed debate on how the food safety of GMO crops should be assessed and regulated is still at an early stage.

5. Regulation of the food use of genetically modified crops

Laws and procedures for regulating the purity and wholesomeness of foods are well established in the developed countries. The regulatory system governing food use of plant products will have profound implications for the application of plant biotechnology. In the USA regulations for food use of GMO crops are made by the Food and Drug Administration (FDA). In the European Community, such regulations are currently the responsibility of the member states. The main issues concern the point in the chain at which regulations will be applied, the decision-making process, and consumer acceptance.

5.1 What should be regulated?

Any regulatory system needs to be clear, consistent and practicable. In order to meet these criteria, some argue that regulation should be based on whether or not certain processes have been used. For others this view is illogical and a regulatory system based on the characteristics of a product, rather than on its method of construction, is more appropriate. Thus once again we are faced with the question: should regulation focus on the product or the process? If the latter route is adopted, should regulation be applied at the level of 'genes' which could potentially be approved for use in a wide range of crop varieties, or at the level of individual GMO crop varieties containing particular genes, or at the level of new foods derived from GMO crops. The current concept in Europe seems to be that regulation will be at the point of introduction of 'novel foods' and indeed the EC has drafted a regulation which takes this approach. Such foods could be completely new (for example, mycoproteins for use in vegetarian meals) or foods whose component materials have been produced or altered through biotechnological processes. Although this concept has a superficial attraction (not least to the plant breeder because obtaining regulatory approval would become 'someone else's problem'), there could be many difficulties in this approach for the marketing of GMO varieties. For example, even in the situation where the processors stipulate which varieties of crops should be grown by the farmer, different varieties will be mixed together in the processing chain. If the processor has to keep the GMO variety separate from others and to gain special permission to market foods derived from GMO crops, the processor may decide to ignore the use of that variety completely. In that case the farmer will not grow it. Developing this point, it may be that only if the GMO variety were so popular with the farming community (for example, if it allowed the farmer to reduce input costs by decreasing the use of chemical sprays) that the farmers pressurized the processors to accept the variety, or if the GMO variety allowed a reduction in the raw materials costs of the processor, would there be a situation where the processor reluctantly agreed to obtain regulatory approval for the novel food.

Some argue, therefore, that it is most appropriate to place regulation at the point of development of the first new variety containing a particular novel gene or combination of genes. If that variety were held to be safe for food manufacture, all subsequent food uses of that gene or gene combination would have generic approval. However, this system would place the plant breeders in the new situation of being required to obtain regulatory approval for food use of their varieties. Not surprisingly many plant-breeding companies, especially the smaller ones, would rather that this duty fell either to the biotechnologists who develop novel genes or to the processors. It may be, of course, that regulations are imposed at several levels.

5.2 The decision-making process

Following on from the issue of deciding the point in the processing chain at which regulation should be applied, there is the technical question of how the safety of novel foods should be assessed. Public acceptance of the rigour with which safety assessment is carried out is likely to have a profound effect on consumer acceptance of food biotechnology. The International Food Biotechnology Council (IFBC) has produced a report that attempts to provide a scientific basis to guide regulators in assessing risk (IFBC, 1990). IFBC has suggested a set of 'decision trees' for microbially derived foods and food ingredients and for foods derived from genetically modified plants (for an example of such a decision tree see Table 3.3).

An important principle for the seeds and food industries in developing genetically modified crops for human consumption is that the degree of investigative scrutiny should be appropriate to the specific characteristics of the novel food. Thus varieties in which the food component is unaltered should not be subject to the same intensity of scrutiny as varieties in which the food components have been altered deliberately so as to improve product quality. For example, if oilseed rape were transformed with an insect resistance gene which was expressed only in the leaves so that the properties of the oil and meal were not altered, the regulatory consider-ations should be very different from the case where the fatty acid profile is altered directly through expression of a gene in the developing seed. It seems likely, nevertheless, that the initial GMO varieties will be subjected to extremely detailed food toxicology testing, regardless of this argument. In fact, many consider (for example, the IFBC) that case-by-case analysis in regulating the first GMO foods is 'inevitable and even desirable'. As our technical knowledge develops, general principles for food safety testing should evolve.

5.3 Food labelling

As many GMO crops will not involve direct changes in the food components, it does not seem appropriate for foods derived from GMO crops to be automatically labelled as such. It may be, nevertheless, that an automatic requirement for labelling is imposed. There would be many consequences from such a decision. Would the labelling be for 'genetically modified food' or for 'genetically modified ingredients'? In the future, for example, the consumer might be offered the opportunity to buy a tin of spaghetti bolognese in which the tomato paste has been genetically modified in some way, whereas all of the other ingredients have not. In this situation it might be appropriate for the label to show which of the

Table 3.3. Decision tree for whole foods. Adapted from the International Food Biotechnology Council (1990).

	IF		
	YES		NO
		Go to	
1. Does the product contain genetic material only from organisms that are traditional foods or related non-food species previously used as sources of genetic variation in developing and improving foods by traditional methods of genetic modification?	2		7
2. Are the *significant constituents* in the product only those found in the parents and related species?	3		4
3. Do these *significant constituents* occur within the documented range for the parental traditional foods?	5		5
4. Does the intake of new constituent(s) under intended or reasonably expected conditions of use present *no unacceptable risk?*	6		10
5. Can the intended or reasonably expected conditions of use result ONLY in a pattern of intake of individual constituents that *does not significantly alter* present intake?	6		Safety evaluation of new constituents
6. Are the *significant nutrients* in the product within the expected range for the closely comparable traditional foods the consumption of which the new food will replace?	Accept		Evaluate consequence
7. Is available knowledge and documentation adequate to characterize the introduced genetic material and adequately assure the *lack of an adverse impact* on the final product?	2 and 4		8
8. Are the expression products of the introduced genetic material inherent constituents of foods?	9		Safety evaluation of new constituents

Table 3.3. *Contd.*

9. Are the expression products of the introduced genetic material present at concentrations inherently found in foods?	2 and 4	Safety evaluation of new constituents
10. Can the new constituents be removed or reduced to acceptable levels by processing?	2	Safety evaluation of new constituents and whole foods

ingredients have been genetically modified, but it would surely not be appropriate to have a label which proclaims that this is 'genetically modified food' as this would be the case for only one of the ingredients. One possibility is for a system of labelling analogous to the 'E' numbers for additives which, while not clogging the label with detailed information, allows interested consumers to obtain detailed information about product composition.

In the UK an important precedent has been set by allowing the use of a genetically modified yeast in baking. The modified yeast which contains genes inserted from a sister strain so as to speed up the production of certain enzymes responsible for fermentation has been approved as safe for food use by the UK Advisory Committee on Novel Foods and Processes and by the Food Advisory Committee. Bakers will not be required to label bread made with this yeast as being 'genetically modified' as the new yeast strain has been genetically combined from two similar yeast strains. New legislation may of course change the labelling requirements for this type of product.

5.4 Consumer acceptance of novel foods

Biotechnology raises certain ethical and social issues which are likely to influence acceptance by the consumer of biotechnologically derived products. These issues can be described by three questions: is it safe? is it natural? is it fair? (Straughan, 1989); the last question is particularly appropriate when considering the genetic modification of animals. Some have sought to introduce product regulations on the basis of subjective answers to these questions and the extent to which there is deemed to be a 'social need' for particular products. Others consider that regulation should be based only on the scientific assessment of safety. Once a product has been declared safe, it would then be for the consumer to decide if he/she wanted to buy any particular product. In either case the ethical issues must be addressed by the proponents of biotechnology. There may well be hard lessons to be learned on this from recent experiences with food irradiation and bovine somatotropin (BST). For example, in both Europe and the

USA there has been considerable protest against the proposed use of BST to increase milk production. This in turn has naturally influenced the policies of the supermarkets, some of which have said that at the present time they do not intend to stock milk produced with synthetic BST. If the benefits from crop biotechnology are not explained adequately to the consumer, it is possible that the supermarkets will take the same line with foods derived from genetic modification. Those involved in developing this technology will need to take steps to ensure that this is not the case.

6. The marketing of genetically modified crop varieties

There are several reasons to consider it unlikely that there will be large-scale commercial introductions of genetically modified crop varieties until the mid or late 1990s. These include uncertainties about the IPR and regulatory climate with respect to commercial release of GMO varieties, the need for evaluation of the field performance of GMO crop varieties, the requirement for further breeding and the time taken for multiplication of seed for commercial sale. The first group of GMO varieties is likely to contain those with genes of agronomic value such as herbicide- and insect-resistance where the varieties will be marketed in essentially the same way as conventionally bred varieties. In time, varieties that meet the specific needs of food processors and speciality chemical industries are also likely to be developed, and here the marketing of GMO varieties may become more integrated into the marketing of the finished product.

As discussed in section 1, the commercial rationale of incorporating novel genetic material into crop varieties is to provide characteristics which offer benefits to users and which differentiate those varieties from the competition. Many analysts consider, however, that varietal improvements in agronomic traits such as herbicide-resistance will be 'evolutionary rather than revolutionary' in nature. Any expectation of large increases in seed company profitability in the short to medium term through the use of genetic engineering is likely, therefore, to be confounded. The marketing and sale of GMO varieties with improved agronomic characteristics may, however, provide seed companies with what they might see to be strategic benefits. Large companies may see the introduction of this technology as an opportunity to stimulate further rationalization in the seeds industry, with those companies who are either unable to access this technology or to carry its products through the regulatory system, becoming competitively disadvantaged.

The development of GMO varieties with specific crop quality and compositional attributes may have a more profound effect on marketing strategy in the seeds business. Whereas an agronomic improvement may be superseded by new agrochemicals or the development of resistance in the pest or disease, genetic modifications in quality are likely to be more long-lived. The product lifetime of such a GMO variety is therefore likely to be longer and less dependent on external factors (although plant breeding from such a variety would doubtless continue to give a 'family' of varieties with improved quality characteristics). Such new varieties may be developed in order to satisfy the demands of the food processor or political demands for self-sufficiency. Development of varieties in response to the needs of the food processor (for example high-solids tomatoes for tomato paste) is likely to lead to further vertical integration in the food chain through acquisition or contractual arrangements with plant breeding companies. In turn the farmer may be restricted to prescribed varieties and cropping practices. Thus instead of marketing varieties directly to the farmer, the seeds industry will increasingly market its products to the processors.

Conclusions

Whereas the initial euphoria about the prospects for genetic modification in crop improvement has subsided, scientific progress has continued to a point where this technology is at a threshold between small-scale field trials and general use of GMO crops in agriculture. Genetically modified varieties of cotton, oilseed rape, sugarbeet, potato, tomato, maize, rice and soyabean crops which confer benefits of improved agronomic performance are expected to be developed towards the end of this century. As technical advances continue, the attributes incorporated into varieties through genetic modification are expected to be more complex and more valuable to the seeds industry, farmers and processors. Following on from the initial successes of R & D in generating herbicide-, virus- and insect-resistant plants, the major uncertainties in translating these advances into commercially successful biotechnologically-derived plant varieties are no longer technical. Instead the main uncertainties relate to the scope for proprietary protection and the timescales and costs for gaining regulatory approval; in turn these factors influence judgments about the commercial viability of investment in crop biotechnology.

Despite these uncertainties, genetic modification has great potential to contribute to more sustainable and environmentally sound agricultural systems and to provide renewable sources of raw materials for industry. Plant breeding is thus a 'green' industry. Given this, the scientific community, industry and governments are faced with the challenge of ensuring that the legal protection and regulatory systems are appropriate and that the consumer is offered the choice of products improved through genetic modification. With sufficient information and knowledge, and today's climate of environmental consciousness, there are great expectations that the consumer, whether farmer, processor or ultimately the shopper, will make the 'green' choice.

Acknowledgements

We would like to thank our colleagues at AGC, Doug Gunnary (Nickersons Seeds) and John Duesing (Ciba-Geigy) for their comments.

References

Beier, F.K., Crespi, R.S. and Strauss, J. (1985) *Biotechnology and Patent Protection. An International Review.* Organization for Economic Cooperation and Development, Paris.

Bent, S.A., Schwaab, R.L., Conlin, D.G. and Jeffrey, R.D. (1987) *Intellectual Property Rights in Biotechnology Worldwide.* Stockton Press, New York.

Boyce Thompson Institute for Plant Research (1987) *Regulatory Considerations: Genetically-engineered Plants.* Center for Science Information, San Francisco.

Council of the National Academy of Sciences (1987) *Introduction of Recombinant DNA-engineered Organisms into the Environment: Key Issues.* National Academy Press, Washington, DC.

Department of Agriculture (1991) Proposed guidelines for research involving the planned introduction into the environment of organisms with modified hereditary traits. *Federal Register* **56**, 4134–4151.

Gasser, C.S. and Fraley, R. (1989) Genetically engineering plants for crop improvement. *Science* **244**, 1293–1299.

Hall, T.C., Kemp, J.D., Slightom, J.L. and Sutton, D.W. (1984) European Patent Application 84302533.9; European Patent 0122791.

Ex parte Hibberd (1985) *United States Patent Quarterly* **227**, 443–449.

International Food Biotechnology Council (1990) Biotechnologies and food: assuring the safety of food produced by genetic modification. *Regulatory Toxicology and Pharmacology* **12**, 3, Part 2.

Kidd, G.H. (1985) The new plant genetics: restructuring the global seed industry. In: *The World Biotech Report 1*, Online, London, pp. 311–321.

Lindsey, K. and Jones, M.G.K. (1989) *Plant Biotechnology in Agriculture.* Open University Press, Milton Keynes, UK.

National Research Council (1989) *Field Testing Genetically Modified Organisms: Framework for Decisions.* National Academy Press, Washington, DC.

Office of Science and Technology Policy (1986) Coordinated framework for regulation of biotechnology. *Federal Register* **51**, 23301–23350.

Office of Science and Technology Policy (1990) Principles for federal oversight of biotechnology: planned introduction into the environment of organisms with modified hereditary traits. *Federal Register* **55**, 31118–31121.

Organization for Economic Cooperation and Development (1986) *Recombinant DNA Safety Considerations.* OECD, Paris.

Purchase, H.G. and MacKenzie, D.R. (1990) *Agricultural Biotechnology: Introduction to Field Testing.* Office of Agricultural Technology, US Department of Agriculture.

Royal Commission on Environmental Pollution (1990) *Thirteenth Report: The Release of Genetically Engineered Organisms to the Environment.* HMSO, London.

Straughan, R. (1989) *The Genetic Manipulation of Plants, Animals and Microbes. The Social and Ethical Issues for Consumers: A Discussion Paper.* National Consumer Council, London.

Tiedje, J.M., Colwell, R.K., Grossman, Y.L., Hodson, R.E., Lenski, R.E., Mack, R.N. and Regal, P.J. (1989) The planned introduction of genetically engineered organisms: Ecological considerations and recommendations. *Ecology* **70**, 298–315.

US Congress, Office of Technology Assessment (1988) *New Developments in Biotechnology (3). Field Testing Engineered Organisms: Genetic and Ecological Issues.* OTA, Washington, DC.

Chapter 4
Genetically Engineered Plants for Herbicide Resistance

Philip M. Mullineaux
John Innes Institute, John Innes Centre for Plant Science Research, Colney Lane, Norwich, NR4 7UH, UK

Introduction

The development of herbicide resistance in crop plants using genetic engineering was one of the first applications of this technology to a commercially important objective. It was also a proving ground for many of the methods which are now routine in experiments using plant transformation. There are three reasons for this. First, the mechanisms of herbicide resistance were at least partly understood prior to the first attempts at producing genetically engineered resistance. This arose from the isolation in the laboratory of resistant strains of bacteria, the selection of herbicide-tolerant plant cells in tissue culture and the study of field-selected, herbicide-tolerant weed and crop species. Second, from such studies it was clear that the introduction of a single gene into a plant was likely to result in the desired phenotype (i.e. herbicide tolerance) which could be readily recognized and was well within the scope of the technology existing in the early 1980s. Third, the main supporters of this work and of other areas of plant biotechnology were (and still are) agrochemical companies. Thus, commercial considerations ensured that considerable resources were placed at scientists' disposal.

Although there is a continued debate and considerable doubt about the ethics of producing herbicide-resistant plants for agriculture (Fox, 1990), the scientific value of the work is beyond question and will be described in detail in this chapter. The main landmarks are given below.

1. The development of genetically engineered herbicide resistance demonstrated that it was possible to express genes from one plant species in another unrelated species.

2. Chimeric genes were constructed, consisting of the coding sequences from bacterial or chloroplast genes under the control of transcription and translation signals from plant nuclear or plant viral genes. These artificial genes could be transferred into plants, efficiently expressed and if necessary, their protein products directed to the chloroplast.

3. Herbicide resistance genes are used as selectable marker genes in plant transformation experiments involving the co-transfer of some other unrelated gene.

4. It has been shown that the gene product from one organism (e.g. a bacterium) can substitute for its inactivated counterpart in the transformed plant and allow the biochemical process to carry on apparently as normal.

5. Last but most important, the study of herbicide-resistant plants has provided a major stimulus to the interaction of plant biochemistry, genetics and molecular biology. In the future, the study of herbicide resistance in plants may lead to an understanding of key metabolic processes to allow a rational approach to the design of herbicides analogous to the way drugs are being designed for specific therapeutic needs. A good candidate would be amino acid biosynthesis, for which many existing herbicides are inhibitors (LaRossa and Falco, 1984).

The aim of this chapter is to identify the strategies that plants and their genetic manipulators have adopted to achieve herbicide resistance and to illustrate this by as many case histories as is possible in the space provided.

1. Strategies

Four mechanisms of herbicide resistance or tolerance can be discerned.

1. Overexpression of the protein which is the target for the herbicide. This will cause tolerance of a plant to a given level of the herbicide by ensuring that a proportion of the total amount of target protein synthesized remains unaffected, allowing the cell to survive.

2. Alteration of the site of herbicide action. In this case, mutation of the target protein renders it less able to bind to the herbicide but still capable of carrying out its biochemical function.

3. Introduction by gene transfer techniques of herbicide detoxification genes from bacteria.

4. Herbicide detoxification by plants, for example by glutathione-*S*-transferase. This enzyme occurs in plants (as well as other organisms) and renders xenobiotics ineffective by conjugating them to reduced glutathione.

As will be seen, plant genetic engineers have copied the first two strategies and the third strategy is novel to genetically manipulated plants. Glutathione-*S*-transferase has been extensively studied in maize at the molecular level, but has not yet been exploited in approaches to plant genetic manipulation for herbicide resistance. However, it is included in this chapter because it underlines the complexity of plant detoxification mechanisms compared with those which have been exploited by genetic engineers and offers the potential for plants to be developed with tolerance to a spectrum of xenobiotics, including herbicides and pollutants.

In addition to these intracellular mechanisms of herbicide resistance, tolerance to inhibitors can be achieved by reducing their uptake into the cell (Harvey and Harper, 1982). However, since the underlying mechanisms have received little attention and are likely to be complex and not readily amenable to genetic manipulation, no further discussion of this phenomenon will be presented here.

2. Overexpression of target proteins

2.1 Resistance in cultured plant cells

Early work on cell cultures of soyabean and maize showed that resistance to the growth inhibitor aminopterin was associated with increased dihydro-folate reductase activity (Shimamoto and Nelson, 1981). Similarly, the high levels of urease in some cultured tobacco cell lines were correlated with resistance to acetohydroxamate (Yamaya and Filner, 1981). However, the underlying cause of the elevated enzyme activities was not investigated. In contrast, the selected resistance of some suspension culture cell lines to the herbicide glyphosate (*N*-(phosphonomethyl)glycine; Figure 4.1) has been studied in some depth.

Glyphosate (marketed as Roundup, by Monsanto Company) is a potent broad-spectrum non-selective herbicide. Early studies using *Rhizobium japonicum*, *Escherichia coli* (*E. coli*), *Chlamydomonas*, duckweed and several higher plant cell cultures showed that growth inhibition could be overcome by adding the aromatic amino acids phenylalanine, tyrosine and tryptophan to their culture media. These data indicated that glyphosate acted on an enzyme common to the biosynthesis of these amino acids (Jaworski, 1972; Gresshoff, 1979; Roisch and Lingens, 1980). Subsequently, it was shown that the shikimic acid pathway enzyme 5-enolpyruvylshikimic acid-3-phosphate (EPSP) synthase of *Aerobacter aerogenes* was the target for glyphosate. The herbicide acts as a competitive inhibitor for one of the enzyme's substrates, phosphoenolpyruvate (Figure 4.1; Steinrucken and Amrhein, 1980). Subsequently, the same mode of action of glyphosate on EPSP synthase from higher plants was confirmed (Amrhein *et al.*, 1983; Nafziger *et al.*, 1984). Increasing by 5–17-fold the expression of the *aro*A gene of *E. coli*, coding for EPSP synthase, by

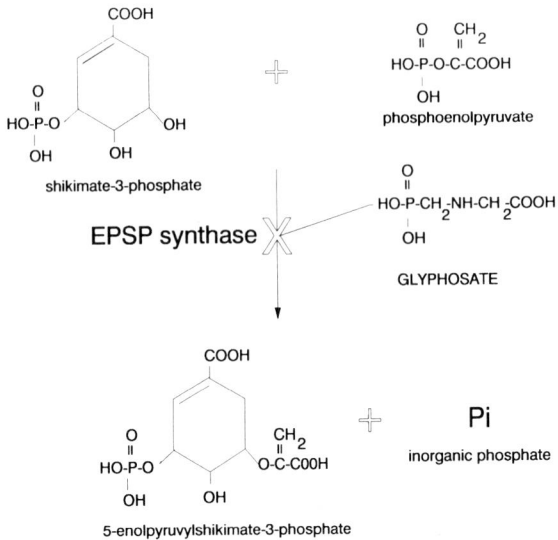

Figure 4.1. Glyphosate and the reaction catalysed by
5-enolpyruvylshikimic acid-3-phosphate (EPSP)
synthase.

placing the gene on a multicopy plasmid resulted in an 8-fold increase in
the minimum inhibitory glyphosate concentration for growth (Rogers *et al.*,
1983). Molecular genetic analysis of glyphosate resistance in *Salmonella
typhimurium* (*S. typhimurium*) and *E. coli* provided yet more evidence
that EPSP synthase was the site of action of glyphosate and is described in
the section on mutation of target proteins.

Selection of plant cell cultures of *Corydalis sempervirens*, carrot and
petunia adapted to grow in medium containing up to 10 mM glyphosate
showed a 12–30-fold increase in EPSP synthase activity. No differences
could be detected either in the rate of uptake of glyphosate or in the
properties of EPSP synthase from the resistant cell lines compared with
those from sensitive cultures (Amrhein *et al.*, 1983; Nafziger *et al.*, 1984;
Steinrucken *et al.*, 1986).

A cDNA coding for EPSP synthase from a glyphosate-tolerant petunia
suspension culture was used as a probe in Northern blots to show that the
EPSP synthase mRNA levels were 20-fold higher in the glyphosate
resistant line (called MP4-G) compared with its parental glyphosate-
sensitive line (called MP4). Subsequently, it was shown that the elevated
levels of EPSP synthase activity and its mRNA in MP4-G were due to
amplification of the EPSP synthase gene to 20 copies, compared with one
copy in MP4 (Shah *et al.*, 1986a).

2.2 Engineering glyphosate resistance by overexpressing an EPSP synthase cDNA

The glyphosate resistance brought about by an overexpression of EPSP synthase in petunia suspension culture cells was reproduced when the same resistance was transferred to transgenic petunia plants (Shah *et al.*, 1986a). This was achieved by placing a cDNA coding for the entire EPSP synthase preprotein under the control of the efficiently expressed cauliflower mosaic virus (CaMV) 35S promoter, which was anticipated to boost the level of intracellular EPSP synthase upon introduction into petunia (Shah *et al.*, 1986a). The 35S-EPSP synthase chimeric gene in a binary Ti vector is shown in Figure 4.2. For *Agrobacterium*-mediated transformation procedures, the reader is referred to Chapter 2 of this volume. Leaf discs

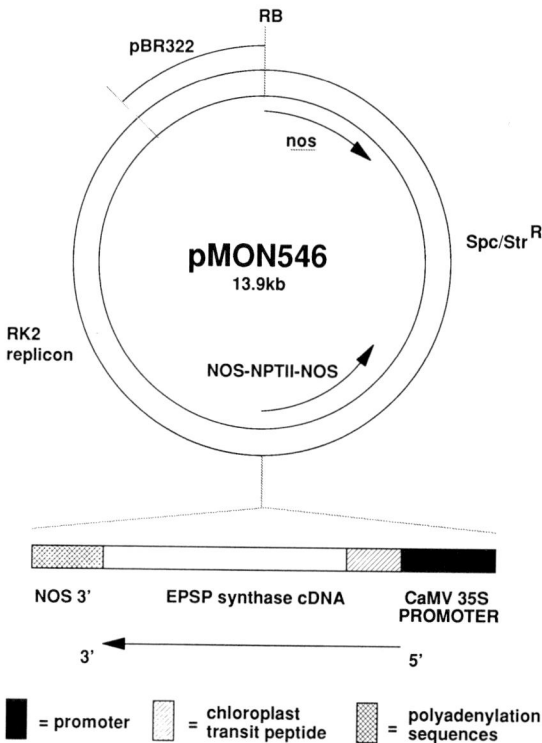

Figure 4.2. The chimeric EPSP synthase gene in the binary Ti vector pMON546 (after Shah *et al.*, 1986a). *nos* is the nopaline synthase gene of T-DNA. Nos-nptII-nos is a chimeric neomycin phosphotransferase gene which confers kanamycin resistance to plant cells.

incubated with *Agrobacterium* containing the chimeric EPSP gene were able to form callus on medium containing 0.5 mM glyphosate whereas control leaf discs failed to do so. EPSP synthase activity in transformed callus was up to 40-fold higher than in control callus, confirming that the engineered EPSP synthase gene could confer glyphosate resistance. Four independent transgenic plants selected for resistance to kanamycin (the resistance gene was co-transferred with the chimeric EPSP synthase gene; see Figure 4.2) could tolerate spraying with Roundup herbicide at a dose equivalent to 0.8 lb per acre, this being two–four times the level required to kill wild type plants.

It should be noted that glyphosate resistance has also been achieved by transferring a mutated bacterial EPSP synthase gene to plants and will be described in the following section.

3. Mutation of target proteins

The second means of achieving resistance is for the target protein to undergo a mutation which reduces its binding to the inhibitor without affecting its function. Genetically engineered resistance to several herbicides has been achieved using genes coding for altered proteins which substitute for their inactivated counterpart in the transgenote.

At the time of writing, four examples of mutated target gene have been used to engineer herbicide tolerance in plants. These are the *aro*A gene from *Salmonella typhimurium* and *E. coli* coding for glyphosate resistance (Comai *et al.*, 1985; della-Cioppa *et al.*, 1987), the *sul*I gene from *E. coli* coding for asulam resistance (Guerineau *et al.*, 1990a), the *psb*A gene of the chloroplast genome of *Amaranthus hybridus* coding for resistance to atrazine (Cheung *et al.*, 1988) and the *csr*l-l (ALS) gene of *Arabidopsis thaliana* coding for resistance to the sulphonyl urea herbicides (Haughn *et al.*, 1988).

In all cases, for the maximum expression of resistance the product of the introduced gene had to be targeted to the chloroplast. To do this, three of the four genes required extensive manipulation, and the resulting constructs are shown in Figure 4.3.

3.1 Glyphosate resistance

In addition to the engineered resistance achieved by overexpressing a wild-type EPSP synthase gene from plants (see previous section and Shah *et al.*, 1986a), use has also been made of the glyphosate-resistant allele of the *aro*A loci of *S. typhimurium* and *E. coli*.

S. typhimurium able to grow on glyphosate-containing medium was

A

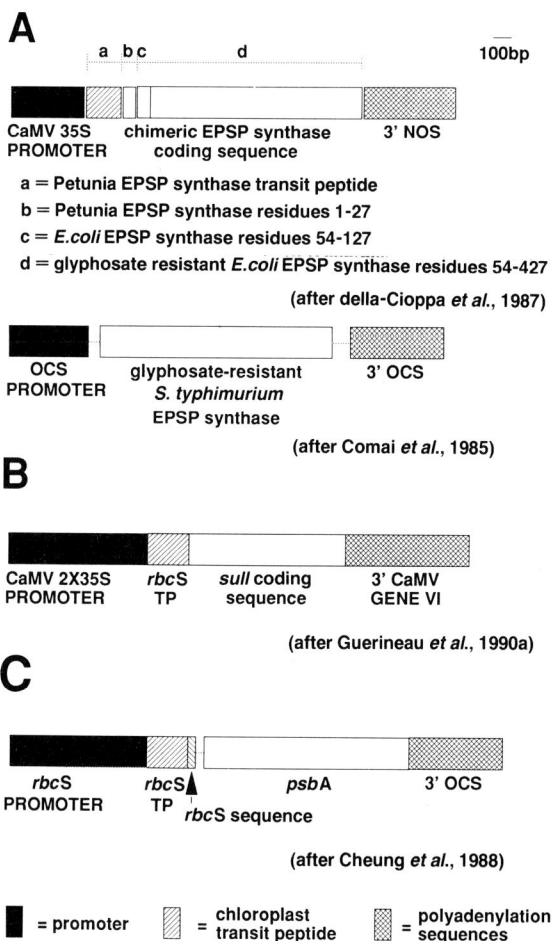

100bp

CaMV 35S   chimeric EPSP synthase   3' NOS
PROMOTER   coding sequence

a = Petunia EPSP synthase transit peptide
b = Petunia EPSP synthase residues 1-27
c = *E.coli* EPSP synthase residues 54-127
d = glyphosate resistant *E.coli* EPSP synthase residues 54-427

(after della-Cioppa *et al.*, 1987)

OCS   glyphosate-resistant   3' OCS
PROMOTER   *S. typhimurium*
EPSP synthase

(after Comai *et al.*, 1985)

B

CaMV 2X35S   *rbc*S   *sull* coding   3' CaMV
PROMOTER   TP   sequence   GENE VI

(after Guerineau *et al.*, 1990a)

C

*rbc*S   *rbc*S   *psb*A   3' OCS
PROMOTER   TP   *rbc*S sequence

(after Cheung *et al.*, 1988)

■ = promoter   ▨ = chloroplast transit peptide   ▩ = polyadenylation sequences

Figure 4.3. Chimeric genes used for engineering resistance to glyphosate (A), asulam (B) and atrazine (C). OCS and NOS are the octopine synthase and nopaline synthase genes respectively of T-DNA. TP is the sequence encoding the chloroplast transit peptide. 2X35S is a CaMV 35S promoter with a double enhancer region.

recovered as spontaneous or chemically induced mutants. A small percentage of these mutations mapped to the *aro*A locus. One of these mutants was shown to synthesize an EPSP synthase with increased resistance to glyphosate (Comai *et al.*, 1983). Enzyme activity was 50% inactivated at 0.01 mM and 1.2 mM glyphosate for the wild-type and resistant EPSP synthases respectively (Stalker *et al.*, 1985). Comparison of the derived amino acid sequences of the EPSP synthases from mutant and wild-type *S. typhimurium* revealed that the substitution of a proline (Pro 101) for a serine residue accounted for the enzyme's decreased affinity for glyphosate (Stalker *et al.*, 1985). A more highly resistant form of EPSP synthase from *E. coli* has also been described (Kishore *et al.*, 1986).

3.2 Engineered glyphosate resistance in plants using the *aro*A gene

Tobacco plants have been engineered for resistance to glyphosate by *Agrobacterium*-mediated introduction of chimeric genes expressing the *aro*A genes of *S. typhimurium* or *E. coli* (Comai *et al.*, 1985; della-Cioppa *et al.*, 1987). The structure of both *aro*A chimeric genes is shown in Figure 4.3A. It should be noted that sequences encoding a chloroplast transit peptide are present only in the construct containing the *E. coli aro*A coding region (Figure 4.3). This difference in the structure of the two chimeric resistance genes may explain why only partial glyphosate tolerance was obtained for the non-targeted *S. typhimurium* gene (Comai *et al.*, 1985). In contrast, the chloroplast-targeted *E. coli* gene product conferred glyphosate resistance on the transgenotes (della-Cioppa *et al.*, 1987). Earlier molecular genetic analysis had shown that an N-terminal 72-amino-acid transit peptide was present on the petunia EPSP synthase preprotein which was required for directing the enzyme to chloroplasts (Shah *et al.*, 1986a; della-Cioppa *et al.*, 1987). In the light of this information, it is perhaps surprising that even partially glyphosate-tolerant tobacco was achieved with the non-targeted chimeric gene product, which presumably remained in the cytosol. This may be explained by the occurrence of complementation across the two cellular compartments by transport of substrates and product of the reaction out of and into the chloroplast respectively (Comai *et al.*, 1985).

3.3 Asulam resistance

Asulam (methyl (4-aminobenzenesulphonyl)-carbamate) is a broad-spectrum herbicide chemically similar to the antibacterial sulphonamide compounds (Figure 4.4). The sulphonamides inhibit the enzyme dihydropteroate synthase (DHPS), an enzyme of the folic acid biosynthetic pathway, by competing with *p*-amino benzoic acid, one of the enzyme's substrates (Franklin and Snow, 1981; Figure 4.4).

Resistance to sulphonamides, including asulam, is conferred on bacteria by various R plasmids. The resistance genes (*sul*) encode a modified DHPS, insensitive to inhibition by sulphonamides (Wise and Abou-Donia, 1985). The nucleotide sequences and expression of the *sul*I genes of R388 and R46 have been determined (Sundstrom *et al.*, 1988; Guerineau and Mullineaux, 1989; Guerineau *et al.*, 1990b). The nature of

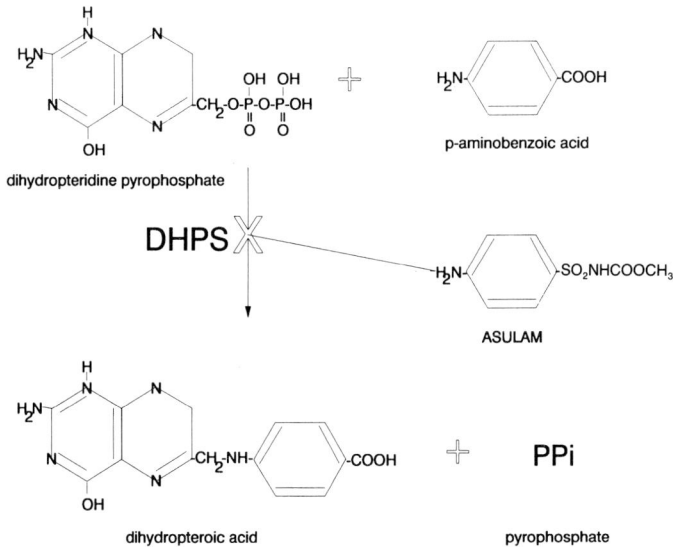

Figure 4.4. Asulam and the reaction catalysed by dihydropteroate synthase (DHPS).

the mutation in DHPS which renders it insensitive to sulphonamides is unknown.

3.4 Engineered asulam resistance in plants

Folic acid derivatives are abundant in chloroplasts and mitochondria (Cossins, 1980), which suggested that the mutated DHPS coded by the *sul*I gene of plasmid R46 would have to be targeted to chloroplasts in order to produce plants resistant to asulam (Guerineau *et al.*, 1990a). To achieve this the coding sequence of the *sul*I gene was fused to the transit peptide sequence (TP) of the pea ribulose bisphosphate carboxylase/oxygenase (Rubisco) gene (*rbc*S). The fusion was made so that the *sul* ATG translation initiation coding sequence was in exactly the same position as the mature Rubisco ATG in the native sequence (Figure 4.5). Thus the bacterial DHPS was expected to be released from the transit peptide in the chloroplast stroma in the same form as synthesized in bacteria, i.e. not as a fusion protein. The Rubisco transit peptide was confirmed as capable of targeting DHPS into chloroplasts. The protein TP-DHPS was synthesized by *in vitro* transcription followed by translation in a wheatgerm lysate. The TP-DHPS was imported into isolated pea chloroplasts and processed to a lower M_r form which had a mobility similar to that of mature-sized DHPS, suggesting that the entire transit peptide had been removed.

For expression in plants, a CaMV 35S promoter with a duplicated enhancer region and a CaMV gene VI polyadenylation signal were placed at the 5′ and 3′ ends respectively of the TP-*sul*I coding sequence (Figure 4.3B). For plant transformation, the chimeric *sul*I gene was inserted into the T-DNA of the binary vector pBIN19 (Bevan, 1984), which also carried the β-glucuronidase (*gus*) gene (Jefferson *et al.*, 1987) placed under the control of the 35S promoter. The plasmid was called pJIT119 (Figure 4.5).

In a series of independent transformation experiments in which asulam or sulphadiazine (a sulphonamide drug) were used as the sole selective

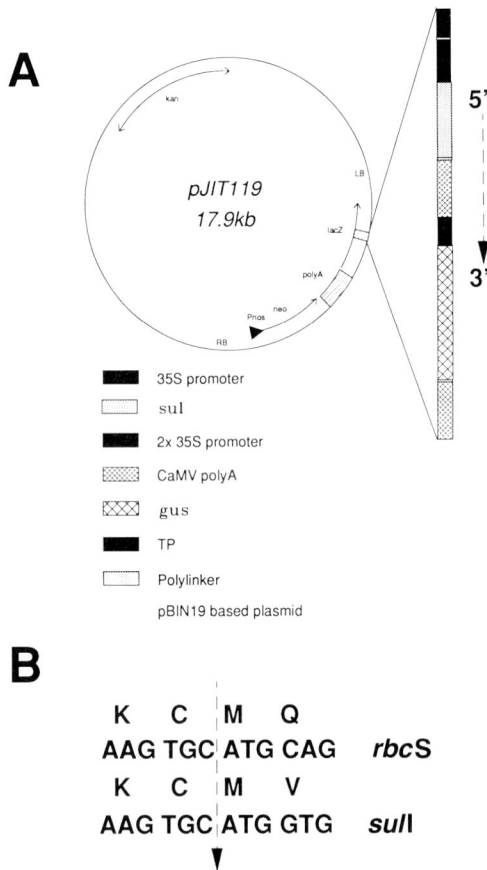

Figure 4.5. A. pJIT119 (after Guerineau *et al.*, 1990a). B. The point of cleavage between the transit peptide and the mature Rubisco protein and the transit peptide-*sul*I coding sequence.

agent, shoots developed on tobacco leaf explants co-cultivated with *Agrobacterium* harbouring pJIT119, but not on those treated with bacteria carrying pJIT59, a pBIN19 derivative carrying the *gus* gene but not the *sul*I gene (Figure 4.6A). Most shoots selected on asulam also rooted on medium containing asulam or sulphadiazine (Figure 4.6B) and showed GUS activity at least 100-fold higher than background levels. Asulam-resistant plants were shown to contain integrated copies of the chimeric *sul*I gene by Southern blot analysis of their genomic DNA. Progeny from five self-fertilized pJIT119 transformants segregated in the ratio three to one in favour of the resistant phenotype, which was consistent with an integration of the T-DNA in one locus. Interestingly, no difference in the growth rate of transformed seedlings, indicating homozygotic or hetero-zygotic status, could be observed on sulphadiazine at 500 mg/l (Figure 4.7). In contrast, GUS assays allowed a clear distinction between seedlings having n or $2n$ alleles of the inserted T-DNA.

It should be noted that the prime objective of the work of Guerineau *et al.* (1990a) was the development of an additional selectable marker gene for plant transformation experiments rather than demonstrating that asulam-resistant plants could be made, since the commercial reason for the latter objective was doubtful. Like the development of glyphosate re-sistance, it may be of more significance that the expression of sulphona-mide resistance in plant tissue, following targeting of the mutated *E. coli* DHPS to chloroplasts, showed that the bacterial enzyme was able to complement the inhibited plant enzyme.

3.5 Atrazine resistance

Atrazine (2-chloro-4-(ethylamino)-6-(isopropylamino)-*s*-triazine; Figure 4.8) is one the *s*-triazine class of herbicides which inhibit photosynthesis. The mechanism of action of atrazine involves the blocking of electron transport at the second stable electron acceptor of photosystem II (Tischer and Strotmann, 1977; Velthuys, 1981). The binding site for the herbicide is a thylakoid membrane polypeptide of the photosystem II complex, known also as the quinone-binding protein (Q_B protein; Gardner, 1981; Mullet and Arntzen, 1981; Pfister *et al.*, 1981). The protein has a molecular size of 32 000 daltons, is tightly bound to the thylakoid membrane and rapidly metabolized and its synthesis is light-regulated (Reisfeld *et al.*, 1978). Translation of the mRNA for Q_B protein occurs on chloroplast ribosomes to produce a precursor polypeptide of 34 500 daltons in maize and spinach (Grebanier *et al.*, 1978; Driesel *et al.*, 1980).

Atrazine-resistant biotypes of many species of weeds have occurred in areas were the herbicide has been intensively used over a long period of time (LeBaron and Gressel, 1982). Triazine resistance is maternally

Philip M. Mullineaux

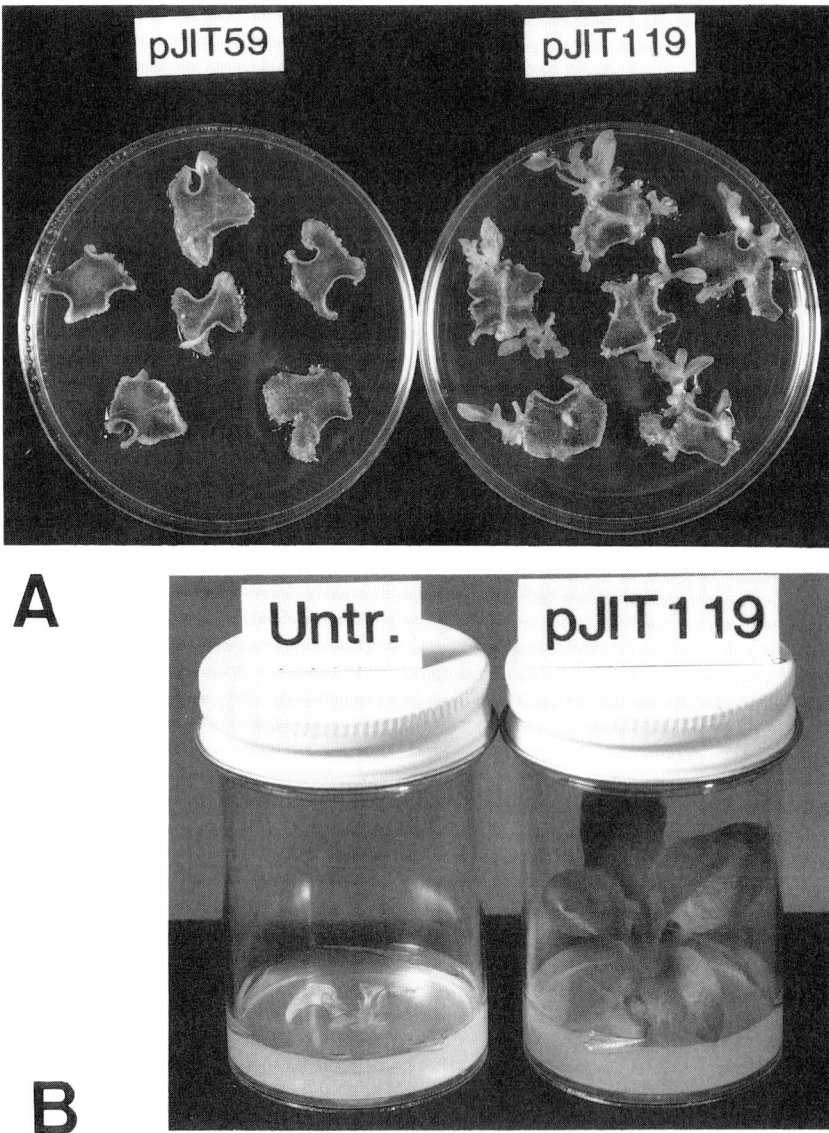

Figure 4.6. A. Co-cultivated tobacco leaf explants after one month on selective medium containing 100 mg/l asulam. B. Tobacco shoots after two weeks on rooting medium containing sulphadiazine at 100 mg/l. Untr. = untransformed shoot.

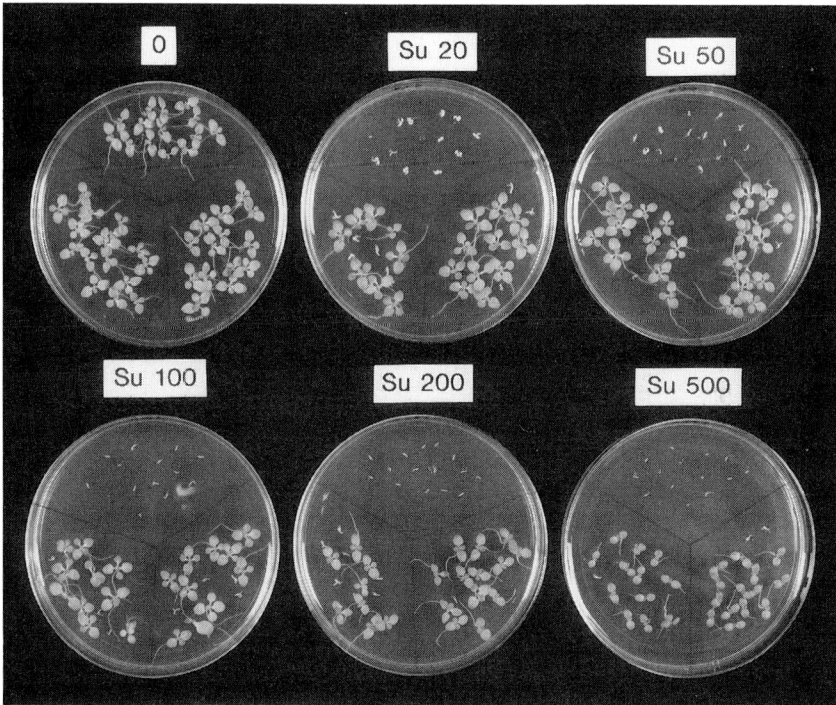

Figure 4.7. Two-week-old tobacco seedlings on medium containing sulphadiazine (SU). The concentrations in mg/l are indicated on the top of each plate. In each plate, the top group of seed was harvested from an untransformed plant and the two bottom groups were harvested from two plants independently transformed with pJIT119.

inheritable and this suggested that the resistance was coded by the chloroplast genome (Souza-Machado *et al.*, 1978). Further evidence was obtained when isolated chloroplasts of an atrazine-resistant biotype of *Amaranthus hybridus* (*A. hybridus*) were shown to contain a 32 000 protein which did not bind the herbicide (Steinback *et al.*, 1981). A comparison of the amino acid sequences of the Q_B protein from atrazine-susceptible and atrazine-resistant biotypes of *A. hybridus*, derived from sequencing their isolated genes (called *psb*A), has revealed that the substitution of a serine (Ser 228) by a glycine was responsible for resistance to the herbicide (Hirschberg and McIntosh, 1983).

In addition to atrazine resistance generated by mutation of the *psb*A gene, detoxification of the herbicide in maize by glutathione-*S*-transferase is an important alternative strategy and is described below.

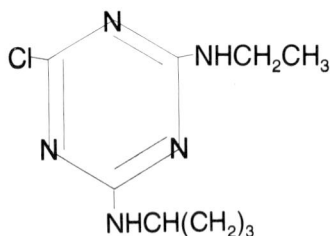

Figure 4.8. Atrazine.

3.6 Engineered atrazine tolerance in tobacco

It has proved possible to transform tobacco with a chimeric gene designed
to express the *psb*A coding sequence of atrazine-resistant *A. hybridus*
(Cheung *et al.*, 1988). It was necessary to construct a chimeric gene
because chloroplast genes were shown not to be expressed from their own
promoters when they were transferred to the nuclear gene in a transformed
plant (Cheung *et al.*, 1988). The chimeric gene consisted of the promoter
and transit peptide sequences of the pea *rbc*S gene (Figure 4.3C). Note
that the construct contains sequences coding for mature Rubisco. This was
done to ensure that the chimeric pre-Rubisco-Q_B hybrid proteins would be
targeted efficiently to the chloroplast. As a result of the manipulation of the
*rbc*S–*psb*A fusion, the mature *A. hybridus* Q_B protein in the chloroplasts
of transgenic tobacco was 3–5 kDa larger than the endogenous tobacco Q_B
protein.

Shoots of control tobacco plants or those transformed with a chimeric
nos promoter mutant *psb*A gene (i.e. a nuclear atrazine-resistant *psb*A
gene lacking a chloroplast transit peptide) bleached after 2 weeks on
rooting medium containing 0.1 mM atrazine. Those plants transformed
with the *rbc*S–*psb*A chimeric genes showed increased tolerance to atrazine, remaining green for a further 3–7 days on average. Many transformants bleached at the same rate as the controls, while, in contrast, one
individual rooted and survived a further 3 weeks before succumbing to
continual exposure to the herbicide (Cheung *et al.*, 1988).

Several possibilities exist for these transgenic plants being only
marginally more tolerant to atrazine than the control plants and less than
the naturally occurring biotype of *Amaranthus* (Cheung *et al.*, 1988).
However, the explanation favoured by the researchers is that only a small
percentage of the photosystem II reaction centres contained atrazine-
resistant Q_B protein, so that on being challenged by the herbicide the
transformed tobacco plant would still have photosynthesis substantially
inhibited. Bleaching would still be caused by the inhibited photoreaction

centres continuing to absorb light. This view was supported by fluorescence studies and by oxygen evolution measurements of photosynthetic electron transport by thylakoid membranes isolated from atrazine-tolerant plants which showed reproducible but low resistance (20–30%) to the herbicide when compared with thylakoid membrane preparations from untrans-formed tobacco (Cheung *et al.*, 1988).

3.7 Resistance to sulphonyl urea herbicides

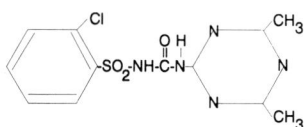

Chlorsulphuron and sulphometuron methyl (Figure 4.9) are both highly potent sulphonyl urea herbicides, sold by the DuPont Company under the trade names Glean and Oust respectively. Sulphometuron methyl is a broad-spectrum non-selective herbicide, while, interestingly, chlorsul-phuron is selective and can be applied to cereal crops without detriment to their growth (Sweetser *et al.*, 1982; LaRossa and Falco, 1984).

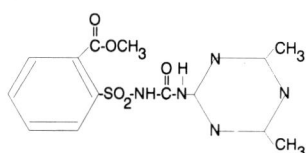

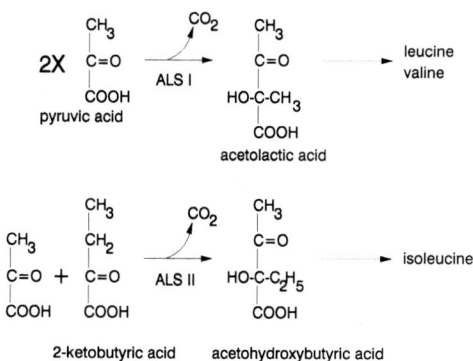

Figure 4.9. Chlorsulphuron and sulphometuron methyl and the reactions catalysed by the isozymes of acetolactate synthase (ALS).

The site of action of the sulphonyl urea herbicides was first studied in bacteria (LaRossa and Schloss, 1984). Sulphometuron methyl was shown to inhibit the growth of *S. typhimurium* on minimal medium only in the presence of L-valine. The inhibition could be overcome by providing L-isoleucine in the growth medium.

Mutations that conferred resistance to sulphometuron methyl on *S. typhimurium* all mapped to the *ilv*G locus. Earlier work with an *S. typhimurium ilv*G mutant showed that it was an auxotroph which could grow on minimal medium supplied with L-isoleucine and either pantothenate or methionine. The latter compounds were required because the mutants accumulated large amounts of 2-ketobutyrate, the substrate for acetolactate synthase (ALS), which is toxic due to its interference with pantothenate biosynthesis (Primerano and Burns, 1982).

Mutants resistant to sulphometuron methyl were similar in behaviour to *rel*A mutants of *S. typhimurium*. A *rel*A mutant is unable to achieve a generalized induction of its amino acid biosynthetic operons and thus has low activities of the enzymes of branched-chain amino acid biosynthesis when grown on minimal medium. Growth of *rel*A mutants was inhibited by sulphometuron methyl in the absence of L-valine and could be relieved by a combination of L-isoleucine and either pantothenate or L-methionine, thus mimicking the *ilv*G mutants (LaRossa and Schloss, 1984). Taken together these data clearly pointed to ALS isozyme II as the target for the herbicide (Figure 4.9) and indicated that other ALS isozymes, whose activity is blocked by feedback inhibition by L-valine, are resistant to sulphometuron methyl but unable to substitute for ALS II in isoleucine biosynthesis by *S. typhimurium* (LaRossa and Schloss, 1984). Kinetic studies on ALS II from *S. typhimurium* suggested that sulphometuron methyl acts on an allosteric site, i.e. not at the active site of the enzyme (LaRossa and Schloss, 1984).

Sequence analysis of spontaneously occurring sulphometuron methyl-resistant ALS II genes from yeast and *E. coli* showed that single amino acid substitutions occurred in different positions in the well-conserved N-terminal region of the two proteins (Figure 4.10). Interestingly, the bacterial mutation conferred only partial resistance to sulphometuron methyl and caused reduced levels of ALS activity. In contrast, the yeast enzyme was insensitive to the herbicide and had unaltered levels of activity (Yadav *et al.*, 1986).

Both tobacco and *Arabidopsis* mutants resistant to sulphonyl urea herbicides have been isolated (Chaleff and Ray, 1984; Haughn and Somerville, 1986).

Tobacco plants resistant to either chlorsulphuron or sulphometuron methyl were isolated by selecting resistant callus in tissue culture (Chaleff and Ray, 1984). The callus was derived from haploid tobacco plants. Resistant callus lines were selected as either spontaneous mutants or after

i) The *E.coli* mutant

E.coli ALS II	V F G Y P G G A I M P V Y D A L
mutant *E.coli* ALS II	V F G Y P G G Ⓥ I M P V Y D A L
yeast ALS	V F G Y P G G A I L P V Y D A I

ii) The yeast and *Arabidopsis* mutants

E.coli ALS II	T G Q V S A P F I G T E A F Q E
yeast ALS	T G Q V P T S A I G T D A F Q E
mutant yeast ALS	T G Q V Ⓢ T S A I G T D A F Q E
Arabidopsis ALS	T G Q V P R R M I G T D A F Q E
mutant *Arabidopsis* ALS	T G Q V Ⓢ R R M I G T D A F Q E

Figure 4.10. Comparison of the deduced amino acid segments in the vicinity of the sulphonyl urea resistance mutations (after Haughn *et al.*, 1988; Yadav *et al.*, 1986). The boxed residue highlights the changed amino acid.

pretreatment with ethylnitrosourea. After several passages of selected calli on herbicide-containing medium, plants were regenerated from them. Extensive genetic analysis was carried out on progeny of the regenerated plants. In most cases the regenerated plants were homozygous for a single dominant nuclear mutation. Linkage analysis of six mutants identified two unlinked genetic loci both of which defined resistance to either chlorsulphuron or sulphometuron methyl. It is important to note that resistance to the herbicides was maintained through several generations of seedlings and calli derived from them. Mutant plants were resistant to a chlorsulphuron concentration 100 times higher than that of control parental plants. Subsequently, it was shown that ALS enzyme from suspension cultures derived from a tobacco plant homozygous for resistance to sulphometuron methyl was resistant to inhibition by both chlorsulphuron and sulphometuron methyl at > 600-fold and 33-fold higher concentrations respectively, than sensitive ALS. This difference in resistance of the mutated ALS to the two sulphonyl urea herbicides correlated with the degree of resistance of the suspension culture to both compounds. Resistant ALS activity and the resistant phenotype always co-segregated, establishing that ALS was the sole target of both chlorsulphuron and sulphometuron methyl in tobacco (Chaleff and Mauvais, 1984).

Chlorsulphuron-resistant mutants of *Arabidopsis thaliana* have also been isolated by screening for the growth of mutagen-treated seedlings in the presence of the herbicide (Haughn and Somerville, 1986). Whole plants and calli derived from them grew on herbicide-containing medium at concentrations up to 300-fold higher than the level required to inhibit growth of wild-type plants. In contrast to the sulphonyl urea-resistant tobacco, genetic analysis of the chlorsulphuron-resistant *Arabidopsis* showed that the resistance was inherited as a single dominant nuclear mutation, designated *csr*, which mapped to chromosome 3. ALS activity from chlorsulphuron-resistant plants was 1000-fold less susceptible to inhibition by the herbicide than the enzyme prepared from wild-type plants. In addition, chlorsulphuron-resistant ALS was 100-fold more resistant to sulphometuron methyl. These data again suggested that the *csr* locus was a chlorsulphuron-resistant allele of the ALS gene.

Wild-type ALS genes from tobacco and *Arabidopsis* have been cloned using a heterologous probe derived from a yeast ALS gene (Mazur *et al.*, 1987). Sequence comparison showed that the ALS sequences from the two plants, from yeast and from the three *E. coli* isozymes shared considerable homology. The cloned DNA fragments were used to investigate the genomic organization of the ALS genes from the two plants. The tobacco genome contains two sets of ALS genes, consistent with the identification of herbicide resistance at two loci (see above and Chaleff and Ray, 1984). The presence of two loci was assumed to be due to tobacco being an allotetraploid species, formed from the fusion of two diploid progenitors, *Nicotiana sylvestris* and *Nicotiana tomentosiformis*. In contrast to tobacco, only a single ALS gene could be detected from *Arabidopsis* and indicated that all the ALS from chorsulphuron-resistant *Arabidopsis* would be resistant to the herbicide, whereas in tobacco this would not be the case. This view was confirmed when the wild-type ALS gene from *Arabidopsis* was used as a probe to recover the mutant allele from the chlorsulphuron-resistant plant (Haughn *et al.*, 1988). Sequence analysis showed that the substitution of a serine residue for a proline residue was responsible for the resistant phenotype and that it is identical to the mutation in the sulpho-meturon methyl-resistant ALS of yeast (Figure 4.10; Yadav *et al.*, 1986).

3.8 Engineered resistance to chlorsulphuron

The ALS gene from chlorsulphuron-resistant *Arabidopsis* was transformed into tobacco as a 5.8-kb genomic fragment inserted into the T-DNA of a disarmed Ti plasmid (Haughn *et al.*, 1988). Unlike all previous examples, this herbicide-resistance gene required no further manipulation and illustrates an advantage of recovering a resistance gene from a convenient plant such as *Arabidopsis* compared with that from a bacterium. The ALS coding sequence is flanked by its own transcription regulatory sequences and contains a transit peptide for targeting the pre-ALS into chloroplasts (Mazur *et al.*, 1987; Haughn *et al.*, 1988). Transformed plants were selected on kanamycin-containing medium using an *aph*3'II chimeric gene which was co-resident with the *csr* gene in the T-DNA and was present at 2–10 copies per haploid genome.

Callus initiated from leaves of the transformed tobaccos was resistant to at least 100 nM chlorsulphuron, whereas callus initiated from control plants failed to grow at 10 nM. Progeny from transformed plants was able to grow at concentrations of chlorsulphuron up to 1000 times higher than the levels normally inhibitory to the growth of untransformed seedlings. The introduced gene(s) segregated approximately three to one (resistant:sensitive), suggesting that the expressed copies were inserted at a single locus. ALS activity was slightly lower in plants transformed with the *Arabidopsis* gene than in control plants, suggesting that overexpression of ALS did not contribute to chlorsulphuron resistance in the transgenic tobacco. However, the ALS activity from the chlorsulphuron-resistant transgenotes was more resistant to inhibition by the herbicide than the enzyme from control plants.

4. Detoxification genes from bacteria

The introduction of herbicide-detoxifying genes from bacteria into plants is a strategy developed by the genetic engineer. The potential for this system was demonstrated by the use of antibiotic-resistance genes from bacteria coding for detoxifying enzymes to select for transformed plant cells resistant to their corresponding antibiotics (Bevan *et al.*, 1983; van den Elzen *et al.*, 1985). The first use of a detoxifying gene for a herbicide was the development of bialaphos-resistant plants by the transfer of a chimeric gene coding for phosphinothricin acetyl transferase (PAT) from *Streptomyces hygroscopicus* (De Block *et al.*, 1987). In keeping with many drug-resistance genes from *Streptomyces* species, bialaphos is produced by this organism and PAT is normally present to protect the bacterium from its own antibiotic (Thompson *et al.*, 1987). A potentially more widespread application is suggested by the isolation of detoxifying genes coding for enzymes of catabolic pathways for halogenated aromatic compounds. This has been successful for producing engineered resistance to 2,4-dichlorophenoxyacetic acid (2,4-D) and bromoxynil (Stalker *et al.*, 1988b; Streber and Willmitzer, 1989). This work has demonstrated that it is not necessary to express all genes involved in the catabolism of a xenobiotic in order to achieve efficient resistance in a plant. It is sufficient to identify and introduce the gene whose product converts the herbicide to a non-phytotoxic form.

In the cases of engineered bialaphos and 2,4-D resistance using bacterial detoxifying genes, transformed plants appear to be more resistant and phenotypically normal compared with plants selected on herbicide-containing media via tissue culture or mutagenized seed (Swanson and Tomes, 1983; Donn *et al.*, 1984; Estelle and Somerville, 1987).

In the examples examined below, the resistance genes did not have to be targeted to any cell compartment and therefore the herbicides are presumably detoxified in the cytosol. It is interesting to note that, in at least two of the examples, resistances to bialaphos and bromoxynil, the targets of the herbicides are proteins which are located in the chloroplast. Therefore, it may be important in engineering the use of a detoxifying enzyme, to ensure that the enzyme is interposed between the incoming herbicide and the site of action of the herbicide in the cell.

4.1 Engineered bialaphos resistance

Bialaphos (the active agent of the herbicide Herbiace (Meiji Seika)) is synthesized by the soil bacterium *Streptomyces hygroscopicus (S. hygroscopicus)*, which is used for the commercial production of the compound. Bialaphos is a non-selective herbicide which is a tripeptide consisting of two L-alanine residues and an analogue of L-glutamic acid called phosphinothricin (PPT; Thompson *et al.*, 1987; Figure 4.11). PPT is an inhibitor of glutamine synthetase in both plants and bacteria. PPT is chemically synthesized and is the active ingredient of the herbicide Basta (Hoescht AG). In contrast to PPT, bialaphos has little inhibitory activity against glutamine synthetase *in vitro* but is converted in to PPT *in vivo* by intracellular peptidases which remove the alanine residues (Figure 4.11). The target enzyme, glutamine synthetase, catalyses the assimilation of ammonia into L-glutamate, forming L-glutamine (Figure 4.11), and is

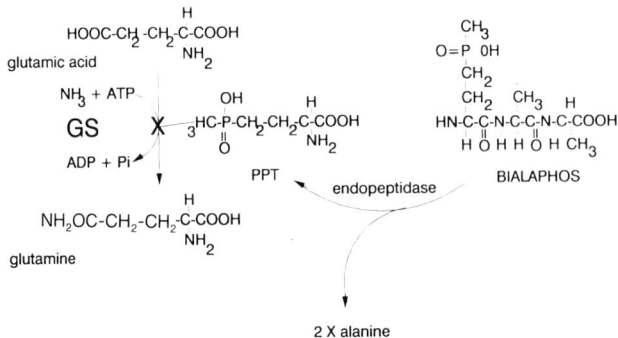

Figure 4.11. Bialaphos and its conversion to L-phosphinothricin (PPT) and the reaction catalysed by the enzyme glutamine synthetase (GS).

central to the regulation of nitrogen metabolism in plants (Miflin and Lea, 1977). The inhibition of glutamine synthetase by PPT causes rapid accumulation of ammonia, leading to cell death (Tachibana *et al.*, 1986).

Streptomyces spp. produce hundreds of antibiotics (Berdy *et al.*, 1980) and have evolved mechanisms to avoid the toxicity of their own products. Such strains often have modifying enzymes which inactivate the antibiotic, and *S. hygroscopicus* is no exception. The bialaphos resistance gene (*bar*) and the genes coding for bialaphos synthesis were shown to be part of an 18-kb gene cluster (Murakami *et al.*, 1986). The gene was subsequently isolated, characterized and shown to encode an acetyltransferase (Figure 4.11; Thompson *et al.*, 1987).

In order to obtain bialaphos resistance in plants, a chimeric gene had to be constructed (De Block *et al.*, 1987). This involved the introduction of an ATG translation initiation codon into the 5′ end of the *bar* coding sequence to replace the GTG initiation codon used in *S. hygroscopicus*. The modified *bar* coding sequence was then placed under the control of a CaMV 35S promoter and the polyadenylation signal of octopine T-DNA gene 7. After insertion into a binary Ti plasmid, the chimeric *bar* gene was transferred via *Agrobacterium*, into plants, PPT-resistant calli were obtained from co-cultivated tobacco mesophyll protoplasts selected on medium containing 50 mg/l PPT as the sole selective agent. All calli tested were resistant to 500 mg/l PPT and most were resistant to 1000 mg/l. Control calli, not transformed with the *bar* gene, were unable to grow on 5 mg/l PPT. The bialaphos resistance gene has recently been used to select transformed maize calli in a procedure analogous to that described for the selection of PPT-resistant tobacco calli (Spencer *et al.*, 1990). The maize calli have subsequently given rise to transformed maize plants (Fromm *et al.*, 1990; Gordon-Kamm *et al.*, 1990).

The *bar* gene was also transformed into tobacco, tomato and potato, using explant culture systems. Interestingly, transformed plants were first selected on kanamycin-containing medium (using a co-resident chimeric *aph*3′II gene) and, once identified, were shown to be resistant to both PPT and bialaphos. In contrast to the selection of PPT-resistant calli, the use of bialaphos as a sole selective agent in transformation of explants may be unsuccessful because of an accumulation of ammonia by untransformed cells poisoning the whole explant (Guerineau *et al.*, 1990a). Nevertheless, *bar*-transformed plants were shown to be highly resistant to both Basta and Herbiace at levels normally used to kill plants. Plants transformed with the chimeric *bar* gene did not accumulate ammonia on spraying with Basta, compared with control plants, which showed an increase of 40-fold in an 8-hour period. Engineered bialaphos resistance displayed typical Mendelian inheritance, behaving as a single dominant trait. Plants were shown to have high steady-state levels of *bar* mRNA, readily immunodetectable

PAT protein and high levels of PPT activity (De Block *et al.*, 1987).

4.2 Engineered bromoxynil resistance

Bromoxynil (3,5-dibromo-4-hydroxybenzonitrile; Figure 4.12) is a potent inhibitor of photosystem II and is sold as Buctril (Rhône-Poulenc). The biochemical target of bromoxynil is known to be localized in the chloroplast and there is evidence to suggest that the herbicide acts by binding to a component of the quinone-binding complex of photosystem II, inhibiting electron transfer (van Rensen, 1982; Sanders and Pallett, 1986). There may also be a low-affinity binding site for bromoxynil in the 32-kDa polypeptide of the same complex (Szigeti *et al.*, 1982; Vermaas *et al.*, 1984).

The herbicide has a short half-life in the soil, since microbes and tolerant plant species can convert the cyano group of bromoxynil to the corresponding amide and acid derivatives (Figure 4.12). The soil bacterium *Klebsiella ozaenae* can convert bromoxynil to 3,5-dibromo-4-hydroxy-benzoic acid, releasing ammonia (McBride *et al.*, 1986). The enzyme which catalyses this reaction is a bromoxynil-specific nitrilase whose gene (*bxn*) is encoded on a plasmid (Stalker and McBride, 1987). The enzyme has been purified to homogeneity (Stalker *et al.*, 1988a). Cloned DNA fragments of the plasmid from *K. ozaenae* could confer bromoxynil resistance on *E. coli*. The sequence of the *bxn* gene was determined and shown to encode a 38-kDa polypeptide (Stalker *et al.*, 1988a).

The *bxn* gene was placed under the control of the tobacco *rbcS* promoter and the T-DNA octopine synthase gene (*ocs*) polyadenylation signal (Stalker *et al.*, 1988a). Transcription directed by the *rbcS* promoter is light-inducible and tissue-specific and was chosen for these experiments because the detoxification of bromoxynil in green tissue was thought to be the minimum requirement for engineering whole-plant resistance. This rationale assumed that the only site of action of bromoxynil was photosynthesis.

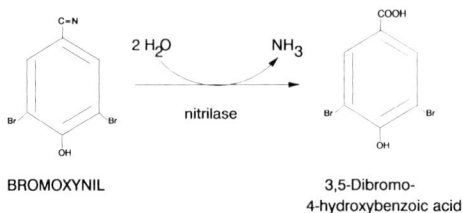

Figure 4.12. Bromoxynil and the reaction catalysed by the bromoxynil-specific nitrilase.

The chimeric *bxn* gene was placed in the T-DNA of a binary Ti vector and transferred into tobacco via *Agrobacterium tumefaciens*. After selection of transformants using a co-transferred kanamycin-resistance marker, the plants were shown to be resistant to bromoxynil up to 10^{-4}M. Control tissue is normally bleached between 10^{-6}M and 10^{-5}M bromoxynil. Resistant plants were shown to have inherited one to three copies of the *bxn* chimeric gene. Like most genes introduced by transformation, bromoxynil resistance was inherited as a single dominant Mendelian trait.

The levels of the nitrilase in the transformants was no greater than 0.01% of total cell protein, although even very low levels (0.0007%) were enough to confer resistance to bromoxynil at commercial levels of application. *In vitro*, nitrilase is only active as a dimer (Stalker *et al.*, 1988a) and therefore this bacterial enzyme must have been correctly assembled *in planta*. Generally, the amount of nitrilase in the transformed plants correlated with their degree of resistance, although the relationship was not linear (Stalker *et al.*, 1988b). In contrast to the levels of the nitrilase protein, *bxn* mRNA was observed at high levels in leaf tissue. No explanation was offered on the discrepancy between the levels of *bxn* mRNA and its protein. As expected, the levels of *bxn* mRNA in stems was lower than in leaves and in roots it was undetectable. These data showed that the strategy and rationale for constructing the *bxn* chimeric gene were correct.

4.3 Engineered resistance to phenoxyacetic acid herbicides

2,4-D (2,4-dichlorophenoxyacetic acid; Figure 4.13) and MCPA (4-chloro-2-methylphenoxyacetic acid) are two examples of the phenoxyacetic acid class of herbicides which show an auxin-like hormone action when applied to dicotyledonous plants. It is thought that the herbicides mimic the function of indole-3-acetic acid, the intrinsic plant auxin. 2,4-D is toxic at high concentrations to dicotyledonous plants but not to monocotyledonous plants and is used as a selective killer of broad-leaved weeds in cereal and grass crops.

The extensive use of these herbicides in agriculture has led to the evolution and widespread occurrence of aquatic and soil bacterial species which possess specific catabolic pathways for the complete degradation of halogenated aromatic compounds (Ghosal *et al.*, 1985). 2,4-D can be metabolized by bacteria of several genera, including *Acinetobacter*, *Alcaligenes*, *Arthrobacter*, *Corynebacterium* and *Pseudomonas* (Streber *et al.*, 1987, and references therein).

In *Alcaligenes eutrophus* JMP134, there are six genes for the degradation of 2,4-D, all located on an 80-kb conjugative plasmid called pJP4 (Don *et al.*, 1985; Streber *et al.*, 1987). The first enzyme of the 2,4-D

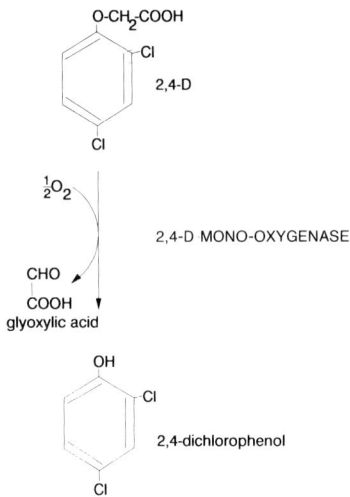

Figure 4.13. 2,4-dichlorophenoxyacetic acid (2,4-D) and the reaction catalysed by the 2,4-D mono-oxygenase.

degradative pathway is a mono-oxygenase encoded by the gene *tfd*A (Figure 4.13). By subcloning ever smaller DNA fragments and expressing 2,4-D resistance in *E. coli*, the *tfd*A gene was isolated, sequenced and shown to encode a 32 kDa polypeptide which displayed a broad substrate specificity, in keeping with the behaviour of this class of enzymes (Streber *et al.*, 1987).

2,4-dichlorophenol, the product of the 2,4-D mono-oxygenase (Figure 4.13), is not toxic to tobacco callus cultures at levels up to 10 mg/l, in contrast to 2,4-D, which severely retards growth at 2 mg/l. This observation suggested that the expression of the *tfd*A gene in plant cells could produce effective resistance to 2,4-D, which indeed proved to be the case (Streber and Willmitzer, 1989).

Engineered resistance to 2,4-D was achieved by constructing a chimeric *tfd*A gene. First, the *tfd*A GTG translation initiation codon was replaced by an ATG codon, required for translation initiation in plants. Then the promoter and polyadenylation sequence were provided by the CaMV 35S promoter and a fragment from the 3′ end of the *ocs* gene respectively (Figure 4.14). The chimeric gene was introduced into tobacco via *Agrobacterium* as a co-integrate with the disarmed Ti plasmid pGV3850kan.

Direct selection of transformants on 2,4-D-containing medium was not possible. Transformed cells could not develop into shoots because they were overgrown or inhibited by untransformed callus, which rapidly developed in the presence of 2,4-D. Furthermore, it was not possible to select for 2,4-D-resistant progeny by germinating seeds on herbicide-containing medium due to the inhibition of root development in the

ATG   TAG

CaMV 35S   DPAM   3' OCS
PROMOTER

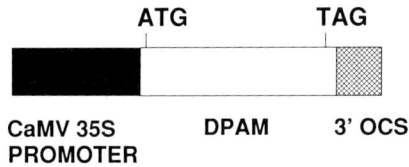

Figure 4.14. The chimeric 2,4-D mono-oxygenase (DPAM) gene (after Streber and Willmitzer, 1989).

seedlings. This may have been caused by poor expression of the chimeric gene in roots or an accumulation of 2,4-D or 2,4-dichlorophenol by transport. However, calli and whole plants selected using a co-resident kanamycin-resistance gene were tolerant to concentrations of 2,4-D 20-fold higher than control material and were phenotypically normal. A *tfd*A-specific mRNA was detected in transformed plants and the chimeric gene was inherited as a single dominant Mendelian character in their progeny.

One important point arising from this work and that which generated bromoxynil-resistant plants is that bacterial detoxification genes coding from one step in a catabolic pathway or the products of their enzymes' reactions need not interfere with the normal metabolism of a plant cell. This may be of some importance in the future in engineering plants which are capable of growing in polluted environments.

5. A plant detoxification system: glutathione-*S*-transferase

The purpose of this final short section is not to give the reader an extensive review of this family of enzymes, but rather to demonstrate the complexity of plant detoxification pathways and to show that considerable potential remains to be exploited with regard to the genetic manipulation of these processes. It is also salutary to compare these sophisticated endogenous systems with those devised by the genetic engineer.

The best characterized group of enzymes engaged in detoxification of herbicides are the glutathione-*S*-transferases (GSTs) of maize, which have been the subject of a recent extensive review (Timmerman, 1989).

Most of the data concern those GSTs of maize which conjugate reduced glutathione with chloro-*s*-triazine and chloracetinilide herbicides, e.g. atrazine, metalochlor and alachlor (Figure 4.15). Three isoenzymes of GST have been identified which have differing specificities for atrazine and metalachlor. These isoforms have the same molecular weight (26 kDa) but

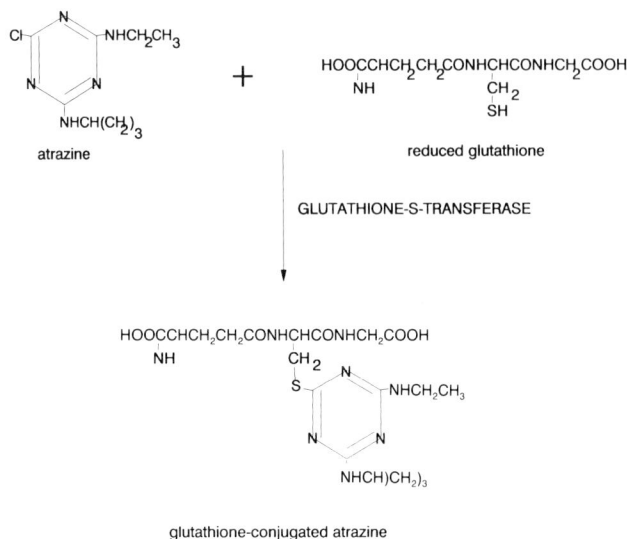

Figure 4.15. Conjugation of reduced glutathione to atrazine by glutathione-*S*-transferase.

have different isoelectric points (Edwards and Owen, 1987). The GST activity responsible for detoxification of alachlor can be resolved into at least three isoenzymes which differ in their molecular weights (29 kDa, 27 kDa and 26 kDa; Mozer *et al.*, 1983; Moore *et al.*, 1986). The three isozymes of maize GST (designated I, II and III) which show activity against alachlor and 1-chloro-2,4-dinitrobenzene (CDNB) do not react with atrazine and similarly the atrazine-GST isoenzymes (of which there are at least two) do not conjugate with CDNB or alachlor (Timmerman, 1989). The active form of all the GST isozymes is a homodimer.

DCNB-GST I and II are constitutively expressed in maize. However, the activity of GST I increases upon treatment of the plant with the herbicide or with safeners. Safeners (also referred to as antidotes) are compounds which induce resistance to herbicides and are not themselves phytotoxic. This induction has been shown to be due to a three- to fourfold induction of transcription of the GST I gene in etiolated maize seedlings (Wiegand *et al.*, 1986) and a ninefold induction in root tissue (Edwards and Owen, 1988). The GST II isoenzyme can only be detected in maize tissue after treatment with safeners and is therefore wholly inducible (Mozer *et al.*, 1983; Moore *et al.*, 1986).

Safeners have also been shown to induce the GST activity which conjugates EPTC-sulphoxide (*S*-ethyl dipropylcarbamothioate-sulphoxide;

Lay and Casida, 1976). Similarly, atrazine-specific GST activity may be induced by intermittent application of the herbicide (Jacchetta and Radosevich, 1981). Interestingly, atrazine pretreatment did not increase tolerance to the herbicides EPTC, alachlor, propachlor and barban, all known to be inactivated by conjugation to glutathione.

Taken together, the cases presented above point to the existence of a GST gene superfamily similar to that in mammals (Lai and Tu, 1986). In addition, other plant species may possess GSTs different from the ones in maize. For example, peas have GST activity which inactivates fluorodifen (2,4′-dinitro-4-trifluoromethyl diphenylether; Frear and Swanson, 1973).

Genes coding for maize GST I and III have been cloned and sequenced. The derived amino acid sequences have at most 45% similarity to mammalian GSTs (Shah *et al.*, 1986b; Grove *et al.*, 1988). Recently, a 44-amino-acid stretch of maize GST III has been shown to share 66% identity with the equivalent region of the *Drosophila* GST 1-1 enzyme (Toung *et al.*, 1990). No molecular characterization of GST II has been reported.

To my knowledge there is no report in the literature of a GST gene from any organism being transferred to and expressed in plants. It is interesting to note that the group which first characterized the maize GST I gene had also successfully engineered plants for resistance to glyphosate (Shah *et al.*, 1986a, b) and therefore had the expertise and technology to introduce a maize GST gene into a dicotyledonous species had they desired to do so. Exploitation of GST in this manner may require a much greater understanding of the way in which the many isoforms of this enzyme interact with glutathione reductase, which supplies reduced glutathione, and with the pathway of glutathione synthesis.

Finally, pathways for the detoxification of herbicides other than GST are known to exist and may attract more attention in the future. For example, the tolerance of cereals to the sulphonyl urea herbicide, chlorsulphuron (see above), may be due to the hydroxylation of the phenyl ring followed by conjugation with a carbohydrate moiety to create an *O*-glycoside of the herbicide (Sweetser *et al.*, 1982).

References

Amrhein, N., Johanning, D., Schab, J. and Schulz, A. (1983) Biochemical basis for glyphosate-tolerance in a bacterium and a plant tissue culture. *FEBS Letters* **157**, 191–196.

Berdy, J., Aszalos, A., Bostian, M. and McNitt, K.L. (1980) *CRC Handbook of Antibiotic Compounds*, Vols. I–VII. CRC Press, Boca Raton, Florida.

Bevan, M.W. (1984) Binary *Agrobacterium* vectors for plant transformation. *Nucleic Acids Research* **12**, 8711–8721.

Bevan, M.W., Flavell, R.B. and Chilton, M.D. (1983) A chimeric antibiotic resistance gene as a selectable marker for plant cell transformation.*Nature* **304**, 184–187.

Chaleff, R.S. and Mauvais, C.J. (1984) Acetolactate synthase is the site of action of two sulfonylurea herbicides in higher plants. *Science* **224**, 1443–1445.

Chaleff, R.S. and Ray, T.B. (1984) Herbicide-resistant mutants from tobacco cell cultures. *Science* **224**, 1148–1151.

Cheung, A.Y., Bogorad, L., Van Montagu, M. and Schell, J. (1988) Relocating a gene for herbicide tolerance: a chloroplast gene is converted into a nuclear gene. *Proceedings of the National Academy of Sciences (USA)* **85**, 391–395.

Comai, L., Sen, L.C. and Stalker, D.M. (1983) An altered aroA gene product confers resistance to the herbicide glyphosate. *Science* **221**, 370–371.

Comai, L., Facciotti, D., Hiatt, W.R., Thompson, G., Rose, R.E. and Stalker, D.M. (1985) Expression in plants of a mutant *aro*A gene from *Salmonella typhimurium* confers tolerance to glyphosate. *Nature* **317**, 741–744.

Cossins, E.A. (1980) One-carbon metabolism. In *The Biochemistry of Plants – A Comprehensive Treatise*, Vol. II, ed. D.D. Davis. Academic Press, New York, pp. 365–418.

De Block, M., Botterman, J., Vandewiele, M., Dockx, J., Thoen, C., Gossele, V., Movva, N.R., Thompson, C., van Montagu, M. and Leemans, J. (1987) Engineering herbicide resistance in plants by expression of a detoxifying enzyme. *EMBO Journal* **6**, 2513–2518.

della-Cioppa, G., Bauer, S.C., Taylor, M.L., Rochester, D.E., Klein, B.K., Shah, D.M., Fraley, R.T. and Kishore, G.M. (1987) Targeting a herbicide-resistant enzyme from *Escherichia coli* to chloroplasts of higher plants. *Bio/Technology* **5**, 579–584.

Don, R.H., Weightman, A.J., Knackmuss, H.-J. and Timmis, K.N. (1985) Transposon mutagenesis and cloning analysis of the pathways for degradation of 2,4-dichlorophenoxyacetic acid and 3-chlorobenzoate in *Alcaligenes eutrophus* JMP134(pJP4). *Journal of Bacteriology* **161**, 85–90.

Donn, G., Tischer, E., Smith, J.A. and Goodman, H.M. (1984) Herbicide-resistant alfalfa cells: an example of gene amplification in plants. *Journal of Molecular and Applied Genetics* **2**, 621–635.

Driesel, A.J., Speirs, J. and Bohnert, H.J. (1980) Spinach chloroplast mRNA for a 32 000 dalton polypeptide: size and location on the physical map of the chloroplast DNA. *Biochimica et Biophysica Acta* **610**, 297–310.

Edwards, R. and Owen, W.J. (1987) Isoenzymes of glutathione S-transferase in *Zea mays. Biochemical Society Transactions* **15**, 1184.

Edwards, R. and Owen, W.J. (1988) Regulation of glutathione S-transferases of *Zea mays* in plants and cell cultures. *Planta* **175**, 99–106.

Estelle, M.A. and Somerville, C. (1987) Auxin-resistant mutants of *Arabidopsis thaliana* with an altered morphology. *Molecular and General Genetics* **206**, 200–206.

Fox, J.L. (1990) Herbicide-resistant plant efforts condemned. *Bio/Technology* **8**, 392.

Franklin, T.J. and Snow, G.A. (1981) *Biochemistry of Antimicrobial Action*, 3rd edition. Chapman and Hall, London, pp. 139–141.

Frear, D.S. and Swanson, H.R. (1973) Metabolism of substituted diphenylether herbicides in plants. I. Enzymatic cleavage of fluorodifen in peas (*Pisum sativum L.*). *Pesticide Biochemistry and Physiology* **3**, 473–482.

Fromm, M.E., Morrish, F., Armstrong, C., Williams, R., Thomas, J. and Klein, T. (1990) Inheritance and expression of chimeric genes in the progeny of transgenic maize plants. *Bio/Technology* **8**, 833–839.

Gardner, G. (1981) Azidoatriazine: photoaffinity label for the site of triazine herbicide action in pea chloroplasts. *Science* **211**, 937–940.

Ghosal, D., You, L.-S., Chatterjee, D.K. and Chakrabarty, A.M. (1985) Microbial degradation of halogenated compounds. *Science* **228**, 135–142.

Gordon-Kamm, W.J., Spencer, M.T., Mangano, M.L., Adams, T.R., Daines, R.J., Start, W.G., O'Brien, J.V., Chambers, S.A., Adams, Jr., W.R., Willetts, N.G., Rice, T.B., Mackey, C.J., Kreuger, R.W., Kausch, A.P. and Lemaux, P.G. (1990) Transformation of maize cells and regeneration of fertile transgenic plants. *Plant Cell* **2**, 603–618.

Grebanier, A.E., Coen, D.M., Rich, A. and Bogorad, L. (1978) Membrane proteins synthesized but not processed by isolated maize chloroplasts. *Journal of Cell Biology* **78**, 734–746.

Gresshoff, P.M. (1979) Growth inhibition by glyphosate and reversal of its action by phenylalanine and tyrosine. *Australian Journal of Plant Physiology* **6**, 177–185.

Grove, G., Zarlengo, R.P., Timmerman, K.P., Li, N.-Q., Tam, M.F. and Tu, C.-P.D. (1988) Characterization and heterospecific expression of cDNA clones of genes in the maize GSH S-transferase multigene family. *Nucleic Acids Research* **16**, 425–438.

Guerineau, F. and Mullineaux, P. (1989) Nucleotide sequence of the sulfonamide resistance gene from plasmid R46. *Nucleic Acids Research* **17**, 4370.

Guerineau, F., Brooks, L., Meadows, J., Lucy, A., Robinson, C. and Mullineaux, P. (1990a) Sulfonamide resistance gene for plant transformation. *Plant Molecular Biology* **15**, 127–136.

Guerineau, F., Brooks, L. and Mullineaux, P. (1990b) Expression of the sulfonamide resistance gene from plasmid R46. *Plasmid* **23**, 35–41.

Harvey, B.M.R. and Harper, D.B. (1982) Tolerance to bipyridylium herbicides. In *Herbicide Resistance in Plants*, ed. H.M. LeBaron and J. Gressel. Wiley, New York, pp. 215–234.

Haughn, G.W. and Somerville, C. (1986) Sulfonylurea-resistant mutants of *Arabidopsis thaliana*. *Molecular and General Genetics* **204**, 430–434.

Haughn, G.W., Smith, J., Mazur, B. and Somerville, C. (1988) Transformation with a mutant *Arabidopsis* acetolactate synthase gene renders tobacco resistant to sulfonyl urea herbicides. *Molecular and General Genetics* **211**, 266–271.

Hirschberg, J. and McIntosh, L. (1983) Molecular basis of herbicide resistance in *Amaranthus hybridus*. *Science* **222**, 1346–1349.

Jacchetta, J.J. and Radosevich, S.R. (1981) Enhanced degradation of atrazine by corn (*Zea mays*). *Weed Science* **29**, 37–44.

Jaworski, E.G. (1972) Mode of action of N-phosphonomethyl glycine inhibition of aromatic amine acid biosynthesis. *Journal of Agriculture and Food Chemistry* **20**, 1195–1198.

Jefferson, R.A., Kavanagh, T.A. and Bevan, M.W. (1987) GUS fusions: β-glucuronidase as a sensitive and versatile gene fusion marker in higher plants. *EMBO Journal* **6**, 3901–3907.

Kishore, G.M., Brundage, L., Kolk, K., Padgette, S.R., Rochester, D., Huynh, K. and della-Cioppa, G. (1986) Isolation, purification and characterisation of a glyphosate tolerant mutant *E. coli* EPSP synthase. *Federal Proceedings* **45**, 1506.

Lai, H.-C.J. and Tu, C.-P.D. (1986) Rat glutathione S-transferase supergene family. *Journal of Biological Chemistry* **261**, 13793–13799.

LaRossa, R.A. and Falco, S.C. (1984) Amino acid biosynthetic enzymes as targets of herbicide action. *Trends in Biotechnology* **2**, 158–161.

LaRossa, R.A. and Schloss, J.V. (1984) The sulfonylurea herbicide sulfometuron methyl is an extremely potent and selective inhibitor of acetolactate synthase in *Salmonella typhimurium*. *Journal of Biological Chemistry* **259**, 8753–8757.

Lay, M.M. and Casida, J.E. (1976) Dichloracetamide antidotes enhance thiocarbamate sulfoxide detoxification by elevating corn root glutathione content and glutathione S-transferase activity. *Pesticide Biochemistry and Physiology* **6**, 442–456.

LeBaron, H. and Gressel, J. (1982) Summary of accomplishments, conclusions and future needs. In *Herbicide Resistance in Plants*, ed. H.M. LeBaron and J. Gressel, Wiley, New York, pp. 349–362.

McBride, K.E., Kenny, J.W. and Stalker, D.M. (1986) Metabolism of the herbicide bromoxynil by *Klebsiella pneumoniae* sub-species *ozaenae*. *Applied and Environmental Microbiology* **52**, 325–330.

Mazur, B.J., Chui, C.F. and Smith, J.K. (1987) Isolation and characterisation of plant genes coding for acetolactate synthase, the target enzyme for two classes of herbicides. *Plant Physiology* **85**, 1110–1117.

Miflin, B.J. and Lea, P.J. (1977) Amino acid metabolism. *Annual Review of Plant Physiology*, **28**, 299–329.

Moore, R.E., Davies, M.S., O'Connell, K.M., Harding, E.I., Wiegand, R.C. and Tiemeier, D.C. (1986) Cloning and expression of a cDNA encoding a maize glutathione S-transferase in *E. coli*. *Nucleic Acids Research* **14**, 7227–7235.

Mozer, T.J., Tiemeier, D.C. and Jaworski, E.G. (1983) Purification and characterization of corn glutathione S-transferase. *Biochemistry* **22**, 1068–1072.

Mullet, J.E. and Arntzen, C.J. (1981) Identification of a 32–34-kilodalton polypeptide as a herbicide receptor protein in photosystem II. *Biochimica et Biophysica Acta* **635**, 236–248.

Murakami, T., Anzai, H., Imai, S., Satoh, A., Nagaoka, K. and Thompson, C.J. (1986) The bialaphos biosynthetic genes of *Streptomyces hygroscopicus*: molecular cloning and characterization of the gene cluster. *Molecular and General Genetics* **205**, 42–50.

Nafziger, E.D., Widholm, J.M., Steinrucken, H.C. and Killmer, J.L. (1984) Selection and characterization of a carrot cell line tolerant to glyphosate. *Plant Physiology* **76**, 571–574.

Pfister, K., Steinback, K.E., Gardner, G. and Arntzen, C.J. (1981) Photoaffinity labelling of a herbicide receptor protein in chloroplast membranes. *Proceedings of the National Academy of Sciences (USA)* **78**, 981–985.

Primerano, D.A. and Burns, R.O. (1982) Metabolic basis for the isoleucine, pantothenate or methionine requirement of *ilvG* strains of *Salmonella typhimurium*. *Journal of Bacteriology* **150**, 1202–1211.

Reisfeld, A., Jakob, K.M. and Edelman, M. (1978) Characterisation of the 32,000 dalton chloroplast membrane protein II. The molecular weight of chloroplast messenger RNAs translating the precursor to P-32000 and full size RUDP carboxylase large subunit. In *Chloroplast Development,* ed. G. Akoyunoglou and I.H. Agyroudi-Akoyunoglou. Elsevier/North-Holland, Amsterdam, pp. 669–742.

Rogers, S.G., Brand, L.A., Holder, S.B., Sharps, E.S. and Brackin, M.J. (1983) Amplification of the *aro*A gene from *Escherichia coli* results in tolerance to the herbicide glyphosate. *Applied and Environmental Microbiology* **46**, 37–43.

Roisch, U. and Lingens, F. (1980) Einfluff von N-(Phosphonometyl)glycine aus das Wachstum und aus die Enzyme der aromaten Biosynthese von *E. coli. Hoppe-Seyler's Zeitschrift für Physiologische und Chemische* **361**, 1049–1058.

Sanders, G.F. and Pallett, K.E. (1986) Studies into the differential activity of the hydroxybenzonitrite herbicides. I. Photosynthetic inhibition, symptom development and ultrastructural changes in two contrasting species. *Pesticide Biochemistry and Physiology* **26**, 116–127.

Shah, D., Horsch, R.B., Klee, H.J., Kishore, G.M., Winter, J.A., Tumer, N.E., Hironaka, C.M., Sanders, P.R., Gasser, C.S., Aykent, S., Siegel, N.R., Rogers, S.G. and Fraley, R.T. (1986a) Engineering herbicide tolerance in transgenic plants. *Science* **233**, 478–481.

Shah, D.M., Hironaka, C.M., Wiegand, R.C., Harding, E.I., Krivi, G.G. and Tiemeier, D.C. (1986b) Structural analysis of a maize gene coding for glutathione S-transferase involved in herbicide detoxification. *Plant Molecular Biology* **6**, 203–211.

Shimamoto, K. and Nelson, O.E. (1981) Isolation and characterization of aminopterin-resistant cell lines in maize. *Planta* **153**, 436–442.

Souza-Machado, V., Bandeen, J.D., Stephenson, G.R. and Lavigne, P. (1978) Uniparental inheritance of chloroplast atrazine tolerance in *Brassica campestris. Canadian Journal of Plant Science* **58**, 977–981.

Spencer, T.M., Gordon-Kamm, W.J., Daines, R.J., Start, W.G. and Lemaux, P.G. (1990) Bialaphos selection of stable transformants from maize cell culture. *Theoretical and Applied Genetics* **79**, 625–631.

Stalker, D.M. and McBride, K.E. (1987) Cloning and expression in *Escherichia coli* of a *Klebsiella ozaenae* plasmid-borne gene encoding a nitrilase specific for the herbicide bromoxynil. *Journal of Bacteriology* **169**, 955–960.

Stalker, D.M., Hiatt, W.R. and Comai, L. (1985) A single amino acid substitution in the enzyme 5-enolpyruvylshikimate-3-phosphate synthase confers resistance to the herbicide glyphosate. *Journal of Biological Chemistry* **260**, 4724–4728.

Stalker, D.M., Malyj, L.D. and McBride, K.E. (1988a) Purification and properties of a nitrilase specific for the herbicide bromoxynil and corresponding nucleotide sequence analysis of the *bxe* gene. *Journal of Biological Chemistry* **265**, 6310–6314.

Stalker, D.M., McBride, K.E. and Malyj, L.D. (1988b) Herbicide resistance in transgenic plants expressing a bacterial detoxification gene. *Science* **242**, 419–423.

Steinback, K.E., McIntosh, L., Bogorad, L. and Arntzen, C.J. (1981) Identification of the triazine receptor protein as a chloroplast gene product. *Proceedings of the National Academy of Sciences (USA)* **78**, 7463–7467.

Steinrucken, H.C. and Amrhein, N. (1980) The herbicide glyphosate is an inhibitor of 5-enolpyruvylshikimic acid-3-phosphate synthase. *Biochemical and Biophysical Research Communications* **94**, 1207–1212.

Steinrucken, H.C., Schulz, A., Amrhein, N., Porter, C. and Fraley, R.T. (1986) Overproduction of 5-enolpyruvylshikimate-3-phosphate synthase in a glyphosate-tolerant *Petunia hybrida* cell line. *Archives of Biochemistry and Biophysics* **244**, 169–178.

Streber, W.R. and Willmitzer, L. (1989) Transgenic tobacco plants expressing a bacterial detoxifying enzyme are resistant to 2,4-D. *Bio/Technology* **7**, 811–816.

Streber, W.R., Timmis, K.N. and Zenk, M.H. (1987) Analysis, cloning and high-level expression of 2,4-dichlorophenoxyacetate monooxygenase gene *tfd*A of *Alicaligenes eutrophus* JMP134. *Journal of Bacteriology* **169**, 2950–2955.

Sundstrom, L., Radstrom, P., Swedburg, G. and Skold, O. (1988) Site-specific recombination promotes linkage between trimethoprim- and sulfonamide-resistance genes. Sequence characterization of *dhfr*V and *sul*I and a recombination active locus of Tn21. *Molecular and General Genetics* **213**, 191–201.

Swanson, E.B. and Tomes, D.T. (1983) Evaluation of birdsfoot trefoil regenerated plants and their progeny after in vitro selection for 2,4-dichlorophenoxyacetic acid. *Plant Science Letters* **29**, 19–24.

Sweetser, P.B., Schow, G.S. and Hutchison, J.M. (1982) Metabolism of chlorsulfuron by plants: biological basis for selectivity of a new herbicide for cereals. *Pesticide Biochemistry and Physiology* **17**, 18–23.

Szigeti, Z., Toth, E. and Paless, G. (1982) Mode of action of photosynthesis inhibiting 4-hydroxybenzonitriles containing nitro group. *Photosynthesis Research* **3**, 347–356.

Tachibana, K., Watanabe, T., Sekizawa, Y. and Takematsu, T. (1986) Accumulation of ammonia in plants treated with bialaphos. *Journal of Pesticide Science* **11**, 33–37.

Thompson, C.J., Movva, N.R., Tizard, R., Crameri, R., Davies, J.E., Lauwereys, M. and Botterman, J. (1987) Characterization of the herbicide-resistance gene *bar* from *Streptomyces hygroscopicus*. *EMBO Journal* **6**, 2519–2523.

Timmerman, K.P. (1989) Molecular characterization of corn glutathione S-transferase isozymes involved in herbicide detoxification. *Physiologia Plantarum* **77**, 465–471.

Tischer, W. and Strotmann, H. (1977) Relationship between inhibitor binding by chloroplast and inhibition of photosynthetic electron transport. *Biochimica et Biophysica Acta* **460**, 113–125.

Toung, Y.-P.S., Hsieh, T.-S. and Tu, C.-P.D. (1990) *Drosophila* glutathione S-transferase 1-1 shares a region of sequence homology with the maize glutathione S-transferase III. *Proceedings of the National Academy of Sciences (USA)* **87**, 31–35.

van den Elzen, P.J.M., Townsend, J., Lee, K.Y. and Bedbrook, J.R. (1985) A chimeric hygromycin resistance gene as a selectable marker in plant cells. *Plant Molecular Biology* **5**, 299–302.

van Rensen, J.J.S. (1982) Molecular mechanisms of herbicide action near photosystem II. *Physiologia Plantarum* **54**, 515–521.

Velthuys, B.R. (1981) Electron dependent competition between plastoquinone and inhibitors for binding to photosystem II. *FEBS Letters,* **126**, 277–281.

Vermaas, W.F.J., Steinback, K.E. and Arntzen, C.J. (1984) Characterization of chloroplast thylakoid polypeptides in the 32kDa region: polypeptide extraction and protein phosphorylation affect binding of photosystem II-directed herbicides. *Archives of Biochemistry and Biophysics* **231**, 226–232.

Wiegand, R.C., Shah, D.M., Mozer, T.J., Harding, E.I., Diaz-Collier, J., Sanders, C., Jaworski, E.G. and Tiemeier, D.C. (1986) Messenger RNA encoding a glutathione S-transferase responsible for herbicide tolerance in maize is induced in response to safener treatment. *Plant Molecular Biology* **7**, 235–243.

Wise, E.M. and Abou-Donia, M.M. (1985) Sulfonamide resistance mechanism in *Escherichia coli*: R plasmids can determine sulfonamide-resistant dihydro-pteroate synthases. *Proceedings of the National Academy of Sciences (USA)* **72**, 2621–2625.

Yadav, N., McDevitt, R.E., Benard, S. and Falco, S.C. (1986) Single amino acid substitutions in the enzyme acetolactate synthase confer resistance to the herbicide sulfometuron methyl. *Proceedings of the National Academy of Sciences (USA)* **83**, 4418–4422.

Yamaya, T. and Filner, P. (1981) Resistance to acetohydroxamate and adaptive increases in urease in cultured tobacco cells. *Plant Physiology* **67**, 1133–1140.

Chapter 5
Fungal Resistance: The Isolation of a Plant R Gene by Transposon Tagging

H. John Newbury

School of Biological Sciences, University of Birmingham,
PO Box 363, Birmingham, B15 2TT, UK

Introduction

Many of the genes that have been used to transform plants in improvement programmes confer resistance, and this includes resistance to herbicides, to viruses or to insects. In this chapter, we will be examining prospects for the genetic engineering of fungal resistance in plants. The development of techniques for DNA transfer to higher plants continues to extend the range of crop species that can be transformed. Thus it may appear that crop plants can now be improved for any character of our choice. However, in most instances the availability of gene transfer technology has only served to emphasize the lack of information concerning plant genes that confer characters of interest. Clearly, appropriate isolated genes must be available if they are to be transferred, but research in plant molecular biology has lagged behind similar work in many other organisms so that the range of plant genes available for biotechnology programmes is relatively small; furthermore, other aspects of plant biology which underpin molecular investigations are often lacking. This situation is reflected by the fact that a high proportion of the strategies of plant improvement using genetic engineering discussed in this book made use of genes of non-plant origin.

1. Aspects of resistance to fungal infection

Initially, it seems sensible to consider a range of plant:pathogen interactions in order to identify resistance genes that may be of use in a biotechnology programme. Any individual plant is subject to a continual rain of microbial pathogens during its life, yet very few of these pathogens are able to infect that plant. In some cases, the pathogens are of a broad host-range type and cannot infect in the absence of particular physical conditions, such as sustained wet weather or damaged tissues. In many other cases, a pathogen has become specialised so that it can efficiently infect one plant species but cannot infect others. For example, a fungal pathogen that has evolved characteristics (pathogenicity genes) which allow it to infect rice plants will probably be unable to infect a cabbage plant. Aspects of the structure and biochemistry of a cabbage leaf render it unsuitable as a substrate for growth of the rice pathogen. This type of interaction is often stated to involve 'non-host resistance', but it is important to note that this particular form of resistance does not involve a molecular defence response by the non-host; simply being a cabbage is good enough to prevent pathogen invasion.

If we examine interactions between individual plant species and a species (often equivalent to a pathovar by workers using bacterial pathogens) of pathogen which has evolved the ability to infect that plant, we can often define genes that control resistance. At this level of interaction, resistant responses by plants to fungal pathogens have been classified into two types, called horizontal and vertical resistance. Individuals displaying horizontal resistance to a particular fungal pathogen do so to all isolates of that species. It is usually accepted that the level of resistance is expressed equally against all isolates (although this may not always be the case). Horizontal resistance is not complete, but usually moderate in nature, allowing some pathogen invasion and spore production. The genetic basis of horizontal resistance is complex, involving a number of genes.

Individuals displaying vertical resistance differ markedly with respect to these characteristics. A resistant individual will exhibit strong resistance to some isolates of the pathogen but will be completely susceptible to other isolates. Resistance to a particular isolate has been found to be conferred by a single gene and these genes are often called 'major genes'or 'R genes'. On the basis of their interactions with individuals of a host species containing different R genes, isolates of a particular pathogen are classified into races. Hence, vertical resistance is said to be race-specific. Pioneering work by Flor (see Flor, 1958, 1971), using flax and the flax rust pathogen *Melampsora lini*, led him to conclude that not only was the vertical resistance exhibited by an individual plant conferred by a single gene, but that a single gene in the pathogen determined whether or not it was virulent on a particular plant. This concept has become known as the gene-for-gene theory and the interaction pattern it predicts is given in Figure 5.1.

The gene-for-gene theory has found substance during the past few years following the cloning of avirulence genes from some plant-pathogenic bacteria, such as *Pseudomonas syringae* (Staskawicz *et al.*, 1984 1987;

		Plant genes	
		R	r
Pathogen genes	A	Incompatible	Compatible
	a	Compatible	Compatible

Figure 5.1. Prediction of interactions between plants and pathogens governed by major genes. The genes are denoted as follows: R = resistance, r = susceptible, A = avirulence, a = virulence. An incompatible response occurs when the plant is able to recognize the pathogen and a resistant response ensues. In an incompatible reaction, the plant is susceptible to pathogen invasion.

Kobayashi *et al.*, 1989) and *Xanthomonas campestris* (Gabriel *et al.*, 1986; Swanson *et al.*, 1988). The presence of the avirulence gene in the pathogen somehow allows the product of the plant R gene to recognize that race, and a resistant response ensues. As with other recognition events involving plants, such as those between stigma and pollen (Knox, 1984) or algal gametes (Wiese, 1984; Callow *et al.*, 1985), the mechanism by which recognition occurs is not known; however, it is clear that, following recognition, a set of genes, often termed response genes, is induced and it is the products of these genes that form the basis of the plant's resistant response (Figure 5.2).

In some instances, it has been shown that the activities of specific enzymes increase during the period of a resistance response; frequently, it has been demonstrated that this follows an increase in the levels of the mRNAs encoding these enzymes. This is not direct evidence that these enzymes represent part of the plant's defence response, although there is circumstantial evidence that this is the case. The catalytic activities of these enzymes often suggest an antipathogen role. For example, the appearance of chitinase activity probably reflects a capacity for degrading many fungal cell walls; endochitinase also shows lysozyme activity and may act against invading bacteria. β-1,3-Glucanases are capable of attacking carbohydrates in fungal cell walls. A series of enzymes catalysing reactions in the phenylpropanoid pathway have commonly been shown to increase in activity during defence responses. In some cases this occurs in tissue producing phenylpropanoid- or flavonoid-based phytoalexins, or showing

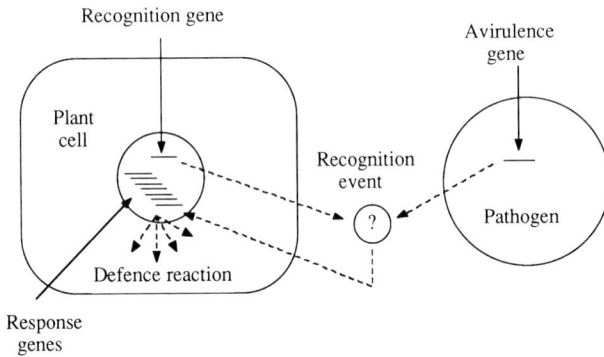

Figure 5.2. Schematic representation of the interaction between a pathogen bearing a particular avirulence gene, and a plant cell, from a potential host species, bearing the complementary major gene for resistance. Following the recognition event, the mechanism of which is not understood, response genes within the plant are induced and the products of these genes act in a defence reaction against the pathogen.

a cell wall lignification response. More detailed information concerning these defence genes and their products is summarized in several excellent reviews (Collinge and Slusarenko, 1987; Dixon *et al.*, 1987; Lamb *et al.*, 1989). Some of the current work in this area is designed to characterize upstream regulatory regions of genes induced during a race-specific resistant response; this will allow the identification of regulatory proteins that bind to these sequences. Analysis of the factors that govern the availability of these proteins for binding to their specific DNA motifs represents a step in the understanding of the proposed signal transduction pathway that operates following an R gene-dependent pathogen recognition event and leads to the induction of response genes.

Plant breeders have expended much effort accumulating as many major genes for resistance to a particular fungal pathogen as possible in modern varieties of important crop plants. These plants therefore show strong resistance to a range of existing races of that pathogen, but are completely susceptible to any new race that may arise. With the enormous numbers of spores produced by fungal pathogens, mutation to novel races is inevitable within a relatively small number of plant generations. Furthermore, the practice of growing large areas of one, or a few, varieties of a particular crop plant (containing a large range of R genes specific to races of that pathogen species) results in a selection against existing races of the pathogen. Thus, in a particular location, a new race can rapidly become established as a dominant form of the pathogen and is an agricultural problem until an appropriate new R gene can be identified in the crop's gene pool and has been introduced into commercial varieties.

Given the background of work on the genetics of plant:pathogen interaction, what are the genetic engineering strategies that could be employed? It seems sensible to ask first whether there are any non-plant genes that can be used to confer fungal resistance. It may, for example, be possible to introduce a constitutively expressed gene, the product of which will inhibit fungal invasion without interfering with plant growth and development; in other words, could a plant be engineered to produce its own systemic fungicide? This approach, which does not appear to have attracted much attention, appears valid, although it would be subject to several obvious criticisms: for instance, it may be possible to apply it only to non-food crops because of the potential dangers of ingesting the fungicide; also, it may rapidly select fungicide-resistant races of fungus and so become ineffective.

Can we instead enhance a resistance mechanism that naturally occurs in plants? A genetic engineering strategy with this aim has a choice of the manipulation of either the so-called response genes or R genes. Many plant genes that are induced during the resistant response to a particular species of fungus have been identified and isolated. One plausible strategy would be to replace the promoters of genes (such as chitinase) with sequences that allow low-level constitutive expression: that is, recognition-independent expression. Thus the plant would be in a permanent low-level resistant reaction. However, as a molecular strategy this approach poses formidable problems. Since it would be extremely difficult to manipulate all the response genes, we would need to identify those whose products are most critical to the production of an effective defence response, and manipulate these. Aside from the size of the task, one could reasonably argue that this approach is fundamentally flawed. Would not the continuous expression of genes involved, for instance, in phenylpropanoid synthesis interfere with the metabolism of the plant and represent a penalty that rendered it unlikely to survive? In other words, if constitutive, low-level expression of antifungal genes is an effective method of defence, would not this system be naturally operating in plants today? But perhaps it is. We do not understand the molecular basis of horizontal resistance, but its main characteristics coincide with those of the plants showing low-level constitutive expression of antifungal genes: their resistance is controlled by multiple genes, is low-level and is apparently not race-specific. The interactions between pathogens and plants showing horizontal resistance have received very little attention, largely because the plant response is weaker and hence more difficult to study. However, work in this area is badly needed, since an understanding of the mechanisms involved in low-level resistance will greatly aid the planning of programmes concerned with the constitutive expression of antifungal genes.

2. Isolation of R genes

An obvious alternative set of target resistance genes for isolation and characterization is the R genes. The justification for endeavours in this area is convincing in academic terms but does not currently attract much interest from biotechnologists. This is because the main advantage of isolating an R gene would lie in the opportunity that is offered to allow the elucidation of the mechanism(s) by which race-specific recognition occurs, and by which response genes are subsequently induced. Until these processes are properly understood, it is not possible to predict the ways in which this information could be used to improve resistance to fungal pathogens. A discussion of genetic engineering programmes is premature until the underpinning molecular pathology studies have been successfully concluded. In spite of this deferral of rewards in the applied field, the molecular characterization of an R gene is commonly regarded as a milestone in plant pathology since it will allow an understanding of the fine detail of gene-for-gene interactions.

So what has been the progress in the isolation of a race-specific fungus resistance gene? Isolation of other plant genes has usually depended upon the availability of certain types of information. For example, gene isolation is facilitated if the specific mRNA is present in high abundance, if the mRNA is unusually large or small, if a short sequence of the gene can be predicted from a partial amino acid sequence of a protein product, or if antiserum to the protein product is available. None of this information is available for R genes and hence 'conventional' gene cloning strategies cannot be employed. There are, however, a number of alternative techniques that can be used in these circumstances. The 'shot-gun transformation' strategy involves the isolation of DNA from a plant variety containing an R gene, and the introduction of random DNA fragments from this donor into a recipient plant which is susceptible to a particular race of a pathogen; transformants are then screened in order to detect an individual which, by virtue of having received the R gene, has become resistant to the appropriate pathogen race. The major problem here, as in most of these strategies, is the enormous size of the typical plant genome. The number of DNA fragments which would have to be transferred in order to randomly introduce the R gene is unmanageably high. This problem is minimized by the use of a plant with an unusually small genome such as *Arabidopsis*, a plant which offers many other advantages as a model in the study of plant molecular genetics (Meyerowitz and Pruitt, 1985). It has been studied extensively for 45 years, using classical genetic techniques, and a series of mutants have been selected or produced. The plant has a generation time of only 5 weeks and can be selfed or crossed with ease, and each individual can produce 10 000 seeds. Furthermore, tens of thousands of plants can be grown in a small room. Unfortunately, *Arabidopsis*:pathogen interactions have, until recently, received little attention.

Another approach which could be used involves the identification of a locus for which a probe is available, an RFLP marker, which has been shown during inheritance studies to be tightly linked to the R gene. The linked locus can then be used as a starting-point for a 'chromosome walking' protocol (Steinmentz *et al.*, 1982; Bender *et al.*, 1983). Such an approach has been suggested for the isolation of the tomato *Tm-2a* gene, which confers resistance to tobacco mosaic virus and has been shown to be tightly linked to five independent genomic clones selected from a large number of DNA fragments by hybridization analyses (Young *et al.*, 1988). By use of a gene library consisting of large DNA fragments resulting from partial digestion with a restriction enzyme, an overlapping series of clones along a chromosome can be identified. Eventually, one of these clones will contain the nearby R locus. Among the problems associated with this technique is the confusion that occurs if repeated-sequence DNA resides between the two defined loci. A further difficulty relates to the correlation between classical chromosomal map units and DNA distances in base pairs. It is not possible to accurately predict how far an apparently tightly linked R gene is from a marker locus in real molecular terms, so that the chromosome walk may become something of a marathon. Some of these difficulties can be avoided if extremely large fragments of DNA containing both the target resistance gene and the linked clones can be prepared. A method has been developed which allows the isolation of DNA fragments larger than 2 million base pairs from tomato protoplasts, using rare-cutting restriction enzymes and pulsed-field gel electrophoresis. For the *Tm-2a* gene, two of the five tightly linked marker clones were found to reside with this resistance gene on a 560-kbp DNA fragment (Ganal *et al.*, 1989). Progress with the isolation of extremely large DNA fragments from *Arabidopsis* has also been made, using pulsed-field electrophoresis; these large fragments have been ligated into a yeast artificial chromosome (YAC) vector and introduced to yeast cells (Guzman and Ecker, 1988). The ability to stably propagate large segments of plant DNA in a simple genetic background should greatly facilitate chromosome walking techniques in species for which extensive RFLP mapping has taken place.

An alternative approach to the problem of isolation of genes that are physically uncharacterized but dominant is to inactivate them using a mobile and recoverable DNA sequence. For this purpose one requires a sequence which: (i) is present, or can be introduced into, the genome containing the target gene; (ii) can become integrated into random positions within the plant genome (some of which will be genes that will become insertionally inactivated); and (iii) has been cloned so that it is available to allow its recovery from the genome, along with flanking

sequences representing the interrupted target gene, using DNA hybridization techniques. At present, two classes of integrating 'element' are
available: the transferred DNA of the *Agrobacterium Ti* or *Ri* plasmids,
and a selection of transposons that have been isolated from two plant
species. In both cases a major constraint, once again, is the size of the plant
genome, which means that large numbers of plants must be screened in
order to detect an individual containing an insertionally inactivated target
gene.

3. Gene cloning using insertional inactivation by T-DNA

The use of *Agrobacterium tumefaciens* as a means of introducing DNA
into plants has received a great deal of attention (see Klee *et al.*, 1987, and
Chapter 2). The technology is based upon the natural ability of this
bacterial plant pathogen to transfer a specific fragment of plasmid DNA
(the T-DNA) into plant cells, where it becomes integrated into the host
genome. *Agrobacterium* has been extensively used to deliver to plants
either genes of agronomic importance or DNA constructs produced in
order to investigate gene regulation and function *in planta*. However, the
introduction of T-DNA sequences into random positions within a genome
also represents a potential for the inactivation of resident genes. In a 'shot-
gun' T-DNA tagging programme using *Arabidopsis*, 136 transformed lines
were used to produce progeny which were screened for segregation of
developmental phenotypes. This has allowed the identification of a single
recessive mutation for dwarfing which co-segregates with a T-DNA insert
(Feldmann *et al.*, 1989) and which is presumably caused by insertional
inactivation of a developmental gene; in another of a series of trans-
formation-induced mutants, an *Arabidopsis* gene for trichome development has been cloned (Marks and Feldmann, 1989).

 The technique is clearly not suitable for species for which *Agrobacterium*-mediated DNA transfer (usually including regeneration of
plants from callus) is not very efficient, because of the large population of
transformants that is required. The process of transformation almost
inevitably involves the regeneration of shoots from disorganized tissue, so
that somaclonal variation is possible within the transformant population. In
the *Arabidopsis* work cited above, it was possible to use a gene transfer
technique involving the infection of germinating seeds with *Agrobacterium*
(Feldmann and Marks, 1987); this particular method does not induce
somaclonal variation.

 Along with the successes of T-DNA tagging of specific developmental
genes, other work suggests that this technique can be improved if certain
modifications to the T-DNA sequence are made (Koncz *et al.*, 1989).

Many T-DNA insertions would be expected to occur in untranscribed regions of the plant genome and hence would not contain an active target gene. To select out these unwanted transformants, the authors introduced a promoterless kanamycin-resistance gene to one end of the T-DNA, so that culture of transformants on this antibiotic selected individuals in which the T-DNA insertion occurred downstream of an active promoter. In fact, about 30% of the T-DNA insertions resulted in kanamycin-resistant transformants suggesting that T-DNA is preferentially integrated into transcriptionally active regions of the genome. A second modification involved the introduction of the pBR322 replicon into the T-DNA construct. This allowed the recovery of T-DNA, and associated flanking sequences representing the inactivated target gene, in the form of a recombinant plasmid that can be directly introduced into *E. coli* following plant DNA isolation. This latter modification would greatly simplify the isolation of target sequences from candidates with tagged genes and is an advantage over the systems to be described for transposon tagging.

4. Transposon-tagging plant genes

Successful programmes employing transposons for plant gene isolation have all used mobile sequences from either maize or antirrhinum. A transposon is a piece of DNA that is capable of changing its position within a genome in an apparently random way. These mobile elements are characterized by the possession of short terminal repeats, which are necessary for their excision from one location and reinsertion into another. An autonomous transposon, such as Activator (Ac) from maize, encodes its own transposase, a product required for mobilizing the element. Autonomous elements may have related non-autonomous transposons which do not encode an active transposase product. Hence, a series of Dissociator (Ds) elements exist in maize lines which, although they are not self-mobilizable, can be mobilized by the transposase of Ac if this element is present within the same plant. Transposons were discovered following the pioneering work on maize by Barbara McClintock (1951, 1965) and appear to exist in all groups of organisms. Plant transposons have been the subject of several detailed reviews (Freeling, 1984; Nevers *et al.*, 1986; Shepherd, 1988). Much of the work on these elements was initiated following observations of variegation in anthocyanin accumulation (in the kernels of maize and the petals of antirrhinum). This occurs when a transposon jumps into and insertionally inactivates a gene involved in anthocyanin synthesis so that initially one cell, and subsequently all the cells produced from it, show a lack of pigment, resulting in pale flecks or streaks on a red background. Conversely, such tissues may exhibit a pale background with red flecks or streaks; in this case an insertionally inactivated gene has become active once more following the departure of a resident transposon to a new location. This reversion phenomenon to produce active genes is a useful characteristic of many plant transposon systems. In this respect these transposons differ from elements such as Tn5 in *E. coli*; here, transposition involves the duplication of the element with one copy staying at the original insertion site and the second moving to a new position within the genome. This mechanism does not allow reversion to a phenotype showing activity of the transposon-inactivated gene. Even with plant transposons excision of the element may result in sufficient disruption to prevent gene expression. Notwithstanding the difficulties associated with the technique, transposon tagging has been employed to isolate a range of plant genes; a list is given in Table 5.1.

The potential of transposon tagging for the isolation of R genes (Bennetzen, 1984) has led to the use of several combinations of plant species, transposons and resistance genes. In some cases, a maize or antirrhinum transposon has been introduced into plant species which have well-characterized R genes, but for which no suitable native transposons exist. This has been achieved by cloning the transposon into a T-DNA region and using *Agrobacterium* to mediate transfer. There is a growing body of evidence to demonstrate that heterologous transposons are capable of movement within a foreign genome. For instance, the maize Ac element has been shown to be mobile in tobacco (Baker *et al.*, 1987), potato (Knapp *et al.*, 1988), tomato (Yoder *et al.*, 1988), *Arabidopsis* (Van Sluys *et al.*, 1987; Schmidt and Willmitzer, 1989) and carrot (Van Sluys *et al.*, 1987). Ac has also been shown to be capable of mobilizing Ds in tobacco (Hehl and Baker, 1989). The maize En-1/Spm element has been shown to

Table 5.1. List of genes isolated by transposon tagging.

Mutable allele	Element	Reference
bz-m2	Ac	Federoff *et al.* (1984); Dooner *et al.* (1985)
al-m(papu)	En	O'Reilly *et al.* (1985)
al-Mum2	Mu	O'Reilly *et al.* (1985)
Pal^rec	Tam3	Martin *et al.* (1985)
c-m668655	En	Paz-Ares *et al.* (1986)
cl-m5	Spm	Cone *et al.* (1986)
c2-ml	Spm	Wienand *et al.* (1986)
bz2-m	Ds	Theres (1986)
P-vv	Mp(Ac)	*Maize Genetic Co-operative Newsletter*(1986)
r-nj:ml	Ac	Dellaporta *et al.* (1988)
O2-m	Spm	Schmidt *et al.* (1987)

transpose in potato (Frey *et al.*, 1989) and tobacco (Pereira and Saedler, 1989), and the antirrhinum Tam3 element is mobile in tobacco (Martin *et al.*, 1989).

Programmes employing heterologous transposons for R gene isolation include that of Ellis and Lawrence (Canberra), who have introduced Ac into flax as a first step in the isolation of a gene conferring race-specific resistance to flax rust. Information on the race-specific interactions between this host and pathogen includes the seminal work of Flor (1958, 1971); there are 30 resistance genes which are arranged in five groups of closely linked or allelic genes. The scheme for mutant isolation has been described by Ellis *et al.* (1988). In Ellis and Lawrence's experiments, where mutations of four different rust-resistance loci were being screened, 1/3000 of the Ac-containing flax progeny showed a change in phenotype to race-specific rust susceptibility, but, so far, all of this change appears to be unrelated to the movement of Ac. Some candidates appear to have lost chromosomal material, and some are haploid. In both cases, the appropriate R gene is presumably missing. Similarly, Jones (Norwich) is introducing the Ac element into tomato with a view to tagging one of the *Fulvia fulva* (*Cladosporium fulvum*) R genes (Dickinson *et al.*, 1990). Work on the Ac element in maize has shown that there is a marked tendency for Ac to move to closely linked positions in the genome rather than long distances. Hence, in order to maximize the chance of insertionally inactivating a target gene, attempts can be made to select transformants in which the T-DNA containing Ac has become inserted close to that gene. This strategy is being used for the isolation of a *Fulvia fulva* resistance gene (Cf2) where the RFLP linkage map available for tomato is being used to monitor the positions of introduced copies of Ac. It is anticipated that the initial screening of the transformants to select those with an appropriately placed Ac element will reduce the size of the population of plants eventually screened for a change in resistance phenotype.

Since plant transposons have only been isolated and appropriately characterized from maize and antirrhinum, R gene-tagging programmes using homologous elements are clearly confined to these two species. The use of maize offers considerable advantages because of the body of work already available concerning the resistance genes of this important crop plant. There are five or six loci specifying maize rust resistance. The Rp1 locus maps at the tip of chromosome 10, and has 14 alleles (Rp1-a to Rp1-m) which can be recognized by the appropriate race/cultivar interaction (Saxena and Hooker, 1968). As with the flax rust resistance loci M and N, detailed genetic analysis has shown that some of the apparent Rp1 alleles in maize are actually closely linked genes (Flor, 1965; Saxena and Hooker, 1968; Mayo and Shepherd, 1980). This is evident from the recovery of reciprocal recombinants between genes Rp1-a, Rp1-c and Rp1-k.

Pryor (Canberra) is using the Ac/Ds transposon system in an attempt to insertionally inactivate one of the rust resistance genes of the Rp1 region. His work in this area has highlighted an important consideration for all transposon-tagging programmes. The initial aim is to obtain a transposon-inactivated gene (here a race-specific susceptible plant) in a screen of many thousand progeny, but it may be that other genetic factors, quite distinct from transposition events, may operate to yield progeny with a change in resistance phenotype. This gene instability has been examined for the Rp1 alleles and the results indicate frequencies of 7/1000 for Rp1-g, 2/15000 for Rp1-d and 0/10000 for Rp1-m (Pryor, 1987a). Similar work by Bennetzen *et al.* (1988) confirms the instability of alleles at the Rp1 locus. It has been proposed that this instability is a consequence of the complexity of the R gene locus, and is due to recombinational events; it has also been suggested that such events may lead to the production of novel resistance genes in plants (Pryor, 1987b). The instability observed means that experiments designed to detect inactivation of most of the Rp1 alleles by transposons would also result in the production of a background of R gene mutants from which the transposon-induced mutations would have to be further selected. In spite of these difficulties, use of the Ac/Ds system has led to the isolation of one candidate plant (from 170000 screened) showing intermediate susceptibility rather than the full resistance conferred by the Rp1-d allele. This appears to be due to Ds insertion. In progeny from this candidate, a line showing full susceptibility has been obtained; in this case, it seems that the initial Ds insertion partly inactivated the gene, but its eventual excision led to complete loss of activity (A. Pryor, personal communication). Molecular analysis of these plants is continuing in an effort to clone the Rp1-d gene.

5. Transposon-tagging an R gene in antirrhinum

Homologous tagging of a fungal-resistance gene in antirrhinum is being carried out by the author and colleagues at Birmingham, UK. The description of this programme is more detailed to allow a greater insight into the problems associated with this area of research. The antirrhinum gene-tagging system has advantages and disadvantages when compared with that available in maize. A range of cloned transposons is available for both plants, but the characteristics of these often differ slightly. For example, a non-autonomous element, such as Ds, can be mobilized by breeding Ac into a maize line, and subsequently any Ds-induced mutations can be stabilized by selecting progeny that do not contain Ac; this facility is not available in antirrhinum. However, use of the antirrhinum element Tam3 has advantages in that not only have experimental lines in which it transposes at high frequency been identified, but the frequency of movement is markedly temperature-dependent; growth of plants at 15 °C results in a 10^3-fold increase in rate of movement compared with growth at 25 °C, allowing external control over element movement.

A clear advantage offered by antirrhinum over maize is the ease with which clonal plant material, such as candidates containing transposon-induced mutations, can be obtained. Conventional cuttings can be taken, but it is also possible to increase the rate of plantlet production using micropropagation (Newbury, 1986). Our recent success with the regeneration of antirrhinum plants from callus (Atkinson *et al.*, 1989) has also led us to explore the possibilities of *Agrobacterium*-mediated transfer of DNA to this species. Initial results suggest that these attempts will be successful; this contrasts with the considerable difficulties encountered by workers attempting to transform cereals such as maize. The opportunity for transformation is important, because strong evidence that a putative R gene has been isolated would come from experiments in which this DNA, upon transfer to a susceptible line, renders it resistant to an appropriate pathogen race.

A major advantage of the maize system lies in the amount of information available regarding pathogen race structures and plant resistance genes. The classical genetic studies that have already been carried out represent an important source of information for both the planning of tagging experiments and the interpretation of subsequent results. There may also be a perceived advantage in working with a major crop plant, since the information obtained can be applied directly to the test species if opportunities for enhancing crop protection occur. However, whilst we do not propose that there is only one mechanism by which R genes may act, we would anticipate homology between an antirrhinum R gene and some of the resistance genes for the same pathogen in other species.

The first stage of our antirrhinum fungal-resistance gene-tagging programme was to select a suitable pathogen with which to work. Our choice of the rust *Puccinia antirrhini* was based upon the knowledge that (i) rusts specific to other plant species had frequently been shown to exhibit a race structure; (ii) the literature indicated that earlier this century a rust-resistance gene had been introduced into many commercial varieties of *A. majus* from the wild species *A. glutinosa*, and that this resistance had subsequently been overcome by mutation in the pathogen (Gawthrop and Jones, 1981); and (iii) the difference between a resistant and a susceptible response was extremely clear, and hence was particularly appropriate for use in the screening of thousands of test plants. This last point is extremely important for any gene-tagging protocol; the large-scale screening procedure will clearly be very difficult if the difference in phenotype scored is of a quantitative nature. Our inoculation involves the transfer of uredio-spores to the leaves of plants at the 6–8-leaf stage, with symptoms developing in susceptible plants within the following 14 days. However, *P.*

antirrhini is not an ideal pathogen for our studies since we cannot culture it *in vitro* but must maintain it on plants. This is time-consuming, and also means that any subsequent attempts to isolate a rust-avirulence gene corresponding to an antirrhinum R gene would face severe difficulties.

Our analysis of *Antirrhinum majus*: *Puccinia antirrhini* interactions involved the challenge of up to 10 varieties of the plant with 55 spore isolates collected from around the world. Challenged plants were maintained separately in an isolation plant propagator; to avoid potential problems associated with mixed race inoculations, much of the work was carried out using spores derived from single pustules. Two races of rust (α and β) have been distinguished (Aitken *et al.*, 1989; see Table 5.2), and again the difference between resistant and susceptible reactions is extremely clear. The results indicate that whilst some varieties show resistance to race α, no resistance to race β has been observed in the plant material available to us.

The results of screening progeny from a large range of pair crosses for foliar resistance to rust race α have indicated that it is inherited in a manner consistent with the involvement of a dominant, nuclear allele at a single locus. In a range of experiments, we have found a 3:1 ratio of resistance to susceptibility in F2 progeny of an original cross between a resistant and a susceptible parent (Aitken *et al.*, 1989). To test whether the apparently single-gene resistance is always due to the same locus, we have performed crosses between different race α-resistant lines and tested progeny for their reaction to race α. The results of these analyses have revealed no evidence for the existence of more than one dominant race α-specific resistance locus. Tests for the possible existence of an independent recessive race α-specific resistance gene have also been made using material which yielded resistance to susceptibility F2 ratios closer to 13:3 than 3:1; however, we have been unable to provide direct evidence for the existence of such a gene from further inheritance studies.

An essential requirement for the R gene-tagging protocol is an *A. majus* genotype which is homozygous for a dominant, race-specific resistance gene, but which is also characterized by a very high rate of movement of transposons; we have termed such genotypes RR(Hitrans). The high rate of movement of Tam3 in such plants is only partially related to the number of copies of the element in the genome (the element is present as 20–30 copies in all of the antirrhinum lines that we employ) but is very clearly affected by genetic background of the individual. The high rate of transposition is important since it has a direct influence on the frequency with which a target resistance gene will be insertionally inactivated; this, in turn, governs the number of plants that must be screened to detect a candidate with a tagged R gene. The *Antirrhinum* group at IPSR (Norwich) gave us the genotype 'line 75', in which the necessary high rate of transposition occurs, but unfortunately this

Table 5.2. A sample of results following the challenge of ten lines of *A. majus* with 16 (out of a total of 55 tested) isolates of *P. antirrhini* urediospores collected from around the world. Codes A, BA, BC, etc., indicate the isolate collection location; R = resistant response, S = susceptible response. (After Aitken *et al.*, 1989.)

Cultivars	Isolates															
	A	BA	BC	I	AJ	O	Y2	Y4	AO	AG	AN	BD	AF	AT	AU	AP
Line 75	S	S	S	S	S	S	S	S	S	S	S	S	S	S	S	S
Avoncroft	S	S	S	S	–	S	S	S	S	S	S	S	S	S	S	S
Victory	S	–	–	S	–	S	S	S	S	S	–	–	S	S	–	S
Trumpet Serenade	S	S	S	S	S	–	–	S	S	S	–	–	–	–	–	–
Crimson Monarch	R	S	S	S	–	S	–	R	S	S	R	S	–	S	S	S
White Monarch	R	S	–	S	S	S	S	R	S	S	R	S	S	S	S	S
Yellow Monarch	R	S	S	–	S	S	S	–	–	–	R	S	–	S	S	–
Orange Glow	R	S	S	S	–	S	S	R	S	S	R	S	S	S	S	S
Wisley Golden Fleece	R	S	S	S	S	S	S	R	S	S	R	S	–	S	S	S
Yellow Freedom	R	S	S	S	S	S	S	R	S	S	R	S	S	S	S	S

genotype does not contain the gene conferring resistance to race α of *Puccinia antirrhini.* Consequently, it was necessary to cross line 75 with commercial varieties to produce initially Rr(Hitrans) progeny, and subsequently, by selfing, to select RR(Hitrans) individuals. Naturally, it was also necessary to perform a further selfing, with selection of parents whose progeny were all resistant, to ensure that particular individuals were homozygous for the R gene.

Caution was required during the production of RR(Hitrans) plants, since crossing can lead to dramatic reduction in transposition rate in progeny. This is due to genetic factors which are, in general, poorly understood, although, in one case, a 'stabilizer' gene which dramatically reduces the rate of Tam3 movement has been identified (Harrison and Fincham, 1968). The simplest method of monitoring the rate of transposon movement in antirrhinum progeny was to assess the frequency of pale flecks on dark-coloured flowers, caused by insertional inactivation of genes involved in anthocyanin synthesis, during the short crossing programme. This rather crude analysis revealed that transposition rates remained high in the plants to be used in our R gene-tagging protocol.

Having produced individual antirrhinum plants that contain two active copies of the R gene in a genome exhibiting high Tam3 movement, there are two basic protocols that can be employed in order to select individuals possessing an insertionally inactivated R gene. In the first, the plants are held at 15 °C to enhance Tam3 movement during gamete formation, and then self-pollinated. A large pool of F1 progeny are grown and, at low frequency, some of these should contain an inactivated copy of the R gene; however, this event will be masked because of the other active copy of this locus. Self-pollination of F1 plants will reveal segregation of susceptibility in F2 progeny of such plants when they are challenged with the pathogen. This strategy has been used effectively elsewhere, but was not employed here because we are interested specifically in tagging an R gene; susceptible candidates could be derived from the protocol described if any gene involved in the resistance mechanism was inactivated and it could prove difficult to distinguish between the classes of candidates obtained. Instead, we have used the protocol described in Figure 5.3. Individuals that were RR(Hitrans) were next held at 15 °C during pollen production, and then crossed with individuals homozygous for the recessive form of the R gene (rr). The bulk of the progeny were, as expected, resistant to race α, being heterozygous at the R locus. Inoculation of the leaves of progeny seedlings at the six-leaf stage has resulted in the identification of a range of candidate plants which exhibit a change in their rust-resistance phenotype. The frequency with which these candidates are being detected is currently 1 in 750. This is much higher than one would expect, given the size of the antirrhinum genome and when compared with the figure of 1 in 8000 reported by the IPSR *Antirrhinum* group following their successful

RR (Hitrans)   Homozygous with respect to
Race α resistance gene.
Also high transposition rate.

15 °C treatment to increase
transposition frequency by
further factor of 1000-fold.

Low temperature-treated   **X**   **r r**   Susceptible to
RR (Hitrans) plants   Race α

Crossed in order to introduce
potentially inactivated R gene into
the germ line, and subsequently
into progeny containing only one
copy of this R gene.

Most **Rr**   i.e. resistant to Race α

Progeny

Rare **R*r** i.e. susceptible to Race α

**Screen progeny to select
candidate plant(s) with
tagged R gene.**

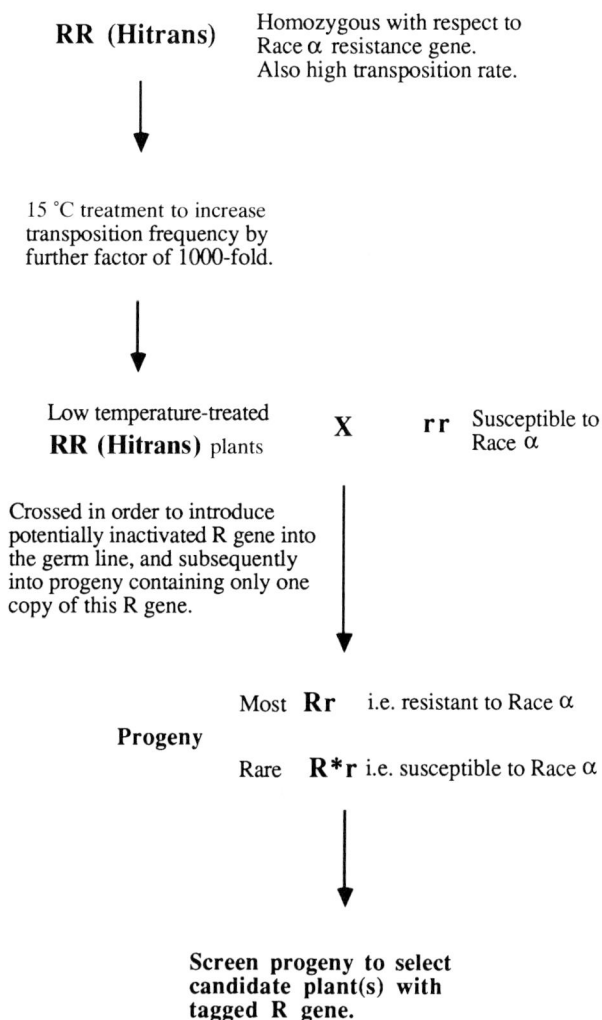

Figure 5.3. The protocol used to select candidate
antirrhinum plants potentially containing a
transposon-tagged race-specific rust-resistance gene.
R = dominant, active, resistance gene; Hitrans = a genetic
background in which the transposon Tam3 moves at an
extremely high rate, which is further enhanced by holding
the plants at 15 °C. Inactivated R genes that enter the male
germ line are selected by using pollen in crosses with female
plants that are susceptible (rr) to rust race α, and screening
the F1 progeny for susceptible candidate plants.

isolation of the *pallida* locus using the same protocol. There are several possible reasons for this. We discount experimental mistakes as an explanation. Although it is possible that contaminated seed or pollen samples could produce these results, it is clear that the floral pigmentation of candidate plants is the same as that of resistant progeny; it is possible to infer from this that the candidate is not the result of accidental self-pollination of the female patient. The spores produced by the rust that infected the candidate plants have been used to inoculate differential *A. majus* cultivars; the results show clearly that the infection was due to spores of race α, and not to contaminating spores of race β.

One explanation for the high frequency of appearance of susceptible candidates is simply that the target R gene is being tagged by Tam3 at a high rate in these plants. The frequency with which a target gene is disrupted by Tam3 appears to be inversely related to the genetic distance between a site occupied by an active copy of the transposon and the gene of interest. It may be that, fortuitously, the R gene lies close to a mobile Tam3 element.

The susceptible plants that we have selected fall into several categories. The first is a chimeric phenotype, in which only one sector of a leaf shows full susceptibility. This is consistent with a somatic transposition event, inactivating the R gene in a single cell and in all cells subsequently derived from it. Whilst of interest, this phenomenon is of no practical use to us because of our inability to regenerate whole plants from mature leaf tissue. A high proportion of our candidate plants show much reduced symptoms, often with pustules concentrated at the leaf periphery. This phenotype is clearly different from that of the bulk of the population, which is pustule-free, but is not a fully susceptible reaction. A small proportion are stunted plants that do not always survive to flowering. Cytological examination has revealed that at least one of these candidates is haploid; on rare occasions it seems the progeny can inherit the haploid genome from the female parent and hence inherit only the recessive form of the R gene. This latter case is a clear example of the production of susceptible plants through genetic events not associated with transposition (as discussed earlier for alleles within the Rp1 locus of maize). The final category of candidates exhibits a clear susceptible reaction and, although any change in resistance phenotype is of interest, these latter are receiving particular attention. Initial tests for most candidates are concerned with the inheritance of the observed phenotype and the frequency with which reversion to full resistance occurs.

Because of the possibility that individuals may lose resistance following genetic events not associated with transposition, we feel that it is important to estimate the frequency of spontaneous mutation at the *A. majus* race α-specific R locus. Our present protocol has been geared to maximizing the movement of Tam3, and hence we have employed a 15 °C treatment. We have been going back to individual RR(Hitrans) plants that have been used to produce the Rr(Hitrans) progeny that we are currently screening, in order to produce further progeny, but this time markedly reducing Tam3 movement by allowing flowering at 25 °C, before crossing with the rr individual used previously. We would expect the frequency of susceptibles in this second screen to be zero if candidates in our first screen were caused by Tam3 movement alone, but to be around 1 in 500 if they were caused by events unconnected with Tam3. This section of our work does not form part of the 'critical path' leading to R gene cloning, but we feel that it is important to learn more about the genetics of the system with which we are dealing.

In part of our previous crossing programme, it became apparent that a floral pigmentation character was co-inherited with the race α-specific resistance locus. A previous publication (Sampson, 1960) had reported that an *A. majus* rust-resistance locus mapped between the *incolorata* and the *eosina* loci. The latter two loci control the ability to produce intermediates in the pathway leading to anthocyanin synthesis. We have reviewed the floral pigment gene status of some of the varieties that we have used to introduce the R gene into our gene-tagging protocol and it is clear that the R gene with which we are working is linked to the *eos* locus.

The availability of a linked marker gene is of great benefit in any tagged protocol. In our case it allows use of a cross as outlined below:

R *eos* (Hitrans)	and	r *Eos*	plants.

R *eos*	r *Eos*
(pink flowered	(dark red-flowered
female parent)	male parent)

Following such crosses, the linked marker will be useful when we reach the stage in our programme of attempting to clone the tagged R gene (r*). We will then want to produce plants in which both copies of the gene are tagged since this facilitates the interpretation of transposon-containing band patterns on autoradiographs. It will not be possible to distinguish between r*r* individuals and r*r or rr individuals by their pathological reaction, since all will be susceptible. However, where we have carried out initial crosses as above, we will be able to distinguish the r*r* genotype on the basis of the associated pink-flowered (*eos/eos*) phenotype.

One of the problems that will face us in the cloning of the R gene from candidate plants which show characteristics of transposon inactivation of this locus, is that the transposon could be one of the other characterized antirrhinum elements and not Tam3. Even though we have increased the rate of movement of Tam3 by 1000-fold by using an appropriate temperature treatment, it is possible that one of the other eight *Antirrhinum* transposons so far identified may occupy the R locus. We expect to be able to distinguish Tam3 inactivation from the effects of other transposons by comparing the rates of reversion to resistance at 25 °C and 15 °C. This, of course, is dependent on obtaining a reasonably high rate of reversion at this locus. In laboratory lines in which Tam3 was inserted at the *nivea* (chalcone synthase) locus the element can have a germinal excision frequency as high as 60% (John Innes Annual Report, 1984). Our recent results indicate that reversion to resistance is apparent in the F1 progeny (following selfing) of six of our initial candidate plants.

The work described so far has employed the techniques of classical genetics and pathology, and has been labour-intensive and time-consuming; the time taken has been a reflection of the generation time of our model plant. However, such basic studies are essential in any tagging protocol to ensure that candidate plants can be selected with confidence before molecular analysis commences. Following the detection of one or more authentic (i.e. with a Tam3-tagged R gene) candidates, with a series of revertants to resistance from each, work will begin to correlate the occurrence of a particular Tam3-bearing restriction fragment, using a Tam3 probe obtained from IPSR, Norwich, in a candidate plant(s) with a loss of resistance to rust race α. The enzyme *Eco*R1 has been commonly used because Tam3 does not contain an internal site for this enzyme. In plants which have reverted to resistance, this copy of Tam3 will have moved into a different position within the genome, and hence a specific *Eco*R1 fragment (which contains the R gene) will disappear and another, of no significance to us, will appear (Figure 5.4). Cloning of the Tam3-tagged R gene-containing fragment from a candidate plant, by cutting this region from an agarose gel, will allow the isolation of DNA fragments flanking the Tam3 element, which should represent parts of the R gene itself. An important factor at this stage of the protocol is that the haploid genome of *A. majus* typically contains 20–30 copies of the Tam3 element. The high number of copies serves to increase the frequency of inactivation of a target gene (and hence to lower the number of progeny that must be screened) although not all of the copies are mobile; some seem to have lost sequences essential for transposition. However, the identification of an appropriate change in banding pattern, following the probing of a filter with a Tam3-specific sequence, may not be easy because of difficulties of band resolution on autoradiographs. Although the technique of correlating the occurrence of a specific Tam3-bearing DNA fragment with a changed phenotype using conventional Southern analysis has previously proved successful (Martin *et al.*, 1985), it has been suggested that the use of the inverse polymerase chain reaction may be useful in this phase of tagging programmes that involve multiple copies of a mobile element. The isolation of a DNA fragment bearing a putative Tam3-inactivated R gene will allow the isolation of the active form of the gene from other antirrhinum individuals as by DNA hybridization. An obvious proof of the authenticity of this gene

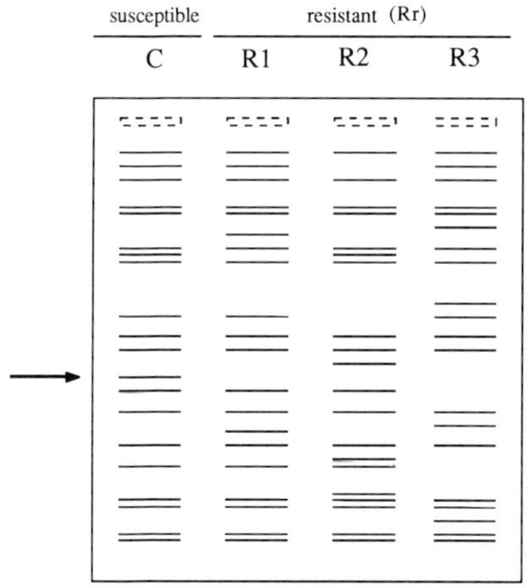

Figure 5.4. A schematic autoradiograph of *Eco*R1-restricted DNA from antirrhinum plants subjected to electrophoresis in an agarose gel, blotted on to a nylon filter, and probed with a radioactive internal fragment of the transposon Tam3. C = DNA from a susceptible candidate plant which contains a tagged R gene (ideally, progeny of the original candidate in which both copies of the R gene are tagged). R1–R3 = DNA from plants which are progeny of the original susceptible candidate, but which exhibit reversion to resistance due to the excision of Tam3 from the R gene; hence, all exhibit the disappearance of the same Tam3-bearing fragment. The Tam3-bearing fragment arrowed in the susceptible candidate track would be cloned, with the expectation that the regions flanking the transposon represent the disrupted R gene.

is to demonstrate that it confers race α-specific resistance to susceptible plants to which it is transferred using an *Agrobacterium* vector (see earlier).

Concluding remarks

A number of strategies for enhancement of fungal resistance by genetic engineering have been described. Almost all of these involve the use of existing plant genes and most depend upon the isolation of an R gene; R gene isolation has previously been reviewed by Ellis *et al.* (1988). The most common approach to R gene isolation makes use of insertional inactivation, usually by a transposon. Our attempts to clone a race-specific antirrhinum rust-resistance gene have been reviewed in order to provide some detailed information regarding stages of such a programme. Transposon-tagging protocols inevitably involve the screening of a large number of progeny to detect a change in race-specific resistance phenotype before molecular analyses can be made, and hence are time-consuming and labour-intensive. However, it seems only a matter of time before an R gene is isolated using these methods.

The isolation of the R locus is the first stage in a process designed to elucidate the mechanism by which plants can recognize specific races of pathogen, and by which they are able to switch on an appropriate defence reaction. Only when these processes have been understood is it possible to define ways in which the acquisition of the gene can contribute to efforts in crop protection. Among the first experiments to be performed with a cloned R gene would be the determination of the structure and cellular location of the R gene product. Answers to these fundamental questions may allow prediction of the mechanism of action of the product, both in recognizing a pathogen race and in triggering the defence reaction of the plant.

Acknowledgements

I would like to acknowledge the work of Liz Aitken and Jim Callow, who are colleagues in the antirrhinum R gene isolation programme. Grateful thanks are also due to Drs Pryor, Ellis and Lawrence (Canberra) for their help in providing unpublished information from their transposon-tagging experiments, and Gordon Todd for his critical review of this manuscript.

References

Aitken, E.A.B., Newbury, H.J. and Callow, J.A. (1989) Races of rust (*Puccinia antirrhini*) of *Antirrhinum majus* and the inheritance of resistance. *Plant Pathology* **38**, 169–175.

Atkinson, N.J., Ford-Lloyd, B.F. and Newbury, H.J. (1989) Regeneration of plants from *Antirrhinum majus* L. callus. *Plant Cell, Tissue and Organ Culture* **17**, 59–70.

Baker, B., Coupland, G., Federoff, N., Starlinger, P. and Schell, J. (1987) Phenotypic assay for excision of the maize controlling element Ac in tobacco. *EMBO Journal* **6**, 1547–1554.

Bender, W., Spierer, P. and Hogness, D.S. (1983) Chromosomal walking and jumping to isolate DNA from the *Ace* and *rosy* loci and the *bithorax* complex from *Drosophila melanogaster*. *Journal of Molecular Biology* **168**, 17–33.

Bennetzen, J.L. (1984) Genetic engineering for improved crop resistance. In *Applications of Genetic Engineering to Crop Improvement*, ed. G.B. Collins and J.G. Petolino. Martinus Nijhoff, Dr W. Junk Publishers, Dordrecht, pp. 491–542.

Bennetzen, J.L., Qin, M.-M., Ingels, S. and Ellingboe, A.H. (1988) Allele-specific and *Mutator*-associated instability at the *Rp1* disease resistance locus of maize. *Nature* **332**, 369–370.

Callow, J.A., Callow, M.E. and Evans, L.V. (1985) Fertilisation in *Fucus*. In *Biology of Fertilisation*, Vol. 2, ed. C.B. Metz and A. Monroy. Academic Press, New York, pp. 389–407.

Collinge, D.B. and Slusarenko, A.J. (1987) Plant gene expression in response to pathogens. *Plant Molecular Biology* **9**, 389–410.

Cone, K.C., Burr, F.A. and Burr, B. (1986) Molecular analysis of the maize anthocyanin regulatory locus C1. *Proceedings of the National Academy of Sciences* **83**, 9631–9635.

Dellaporta, S.L., Greenblatt, I., Kermicle, J., Hicks, J.B. and Wessler, S. (1988) Molecular cloning of the maize *R-nj* allele by transposon tagging with Ac. In *Chromosome Structure and Function*, ed. J.P. Gustafson and R. Appels. Plenum Press, New York, pp. 263–282.

Dickinson, M., Jones, D., Thomas, C., Harrison, K., English, J., Bishop, G., Scofield, S., Hammand-Kosack, K. and Jones, J.D.G. (1990) Strategies for the cloning of genes in tomato for resistance to *Fulvia fulva*. In *Advances in Molecular Genetics of Plant–Microbe Interactions*, ed. H. Hennecke and D.P.S. Verma. Kluwer Academic Publishers, Dordrecht, pp. 276–279.

Dixon, R.A., Bolwell, G.P., Hamdan, M.A.M.S. and Robbins, M.P. (1987) Molecular biology of induced resistance. In *Genetics and Pathogenesis*, ed. P.R. Day and G.J. Jellis, Blackwell Scientific Publications, Oxford, pp. 245–259.

Dooner, H.K., Weck, E., Adams, S., Ralston, E., Favreau, M. and English, J. (1985) A molecular genetic analysis of insertions in the *bronze* locus in maize. *Molecular and General Genetics* **200**, 240–246.

Ellis, J.G., Lawrence, G.J., Peacock, W.J. and Pryor, A.J. (1988) Approaches to cloning plant genes conferring resistance to fungal pathogens. *Annual Review of Phytopathology* **26**, 245–263.

Federoff, N.V., Furtek, D.B. and Nelson, O.E. (1984) Cloning of the bronze locus in maize by a simple and generalizable procedure using the transposable element *Activator* (Ac). *Proceedings of the National Academy of Sciences (USA)* **81**, 3825–3829.

Feldmann, K.A. and Marks, M.D. (1987) *Agrobacterium*-mediated transformation of germinating seeds of *Arabidopsis thaliana*: a non-tissue culture approach. *Molecular and General Genetics* **208**, 1–9.

Feldmann, K.A., Marks, M.D., Christianson, M.L. and Quatrano, R.S. (1989) A dwarf mutant of *Arabidopsis* generated by T-DNA insertion mutagenesis. *Science* **243**, 1351–1354.

Flor, H.H. (1958) The complementary genic systems in flax and flax rust. *Advances in Genetics* **8**, 29–54.

Flor, H.H. (1965) Tests for allelism of rust resistance genes in flax. *Crop Science* **5**, 415–418.

Flor, H.H. (1971) Current status of the gene-for-gene concept. *Annual Review of Plant Phytopathology* **9**, 275–296.

Freeling, M. (1984) Plant transposable elements and insertion sequences. *Annual*

Review of Plant Physiology **35**, 277–298.

Frey, M., Tavantzis, S.M. and Saedler, H. (1989) The maize En-1/Spm element transposes in potato. *Molecular and General Genetics* **217**, 172–177.

Gabriel, D.W., Burges, A. and Lazo, G.R. (1986) Gene-for-gene interactions of five cloned avirulence genes from *Xanthomonas campestris* pv. *malvacearum* with specific resistance genes in cotton. *Proceedings of the National Academy of Science* **83**, 6415–6419.

Ganal, M.W., Young, N.D. and Tanksley, S.D. (1989) Pulsed field gel electrophoresis and physical mapping of large DNA fragments in the *Tm-2a* region of chromosome 9 in tomato. *Molecular and General Genetics* **215**, 395–400.

Gawthrop, F.M. and Jones, B.M.G. (1981) Evidence for genetic change in *Antirrhinum* rust (*Puccinia antirrhini*). *Annals of Botany* **47**, 197–202.

Guzman, P. and Ecker, J.R. (1988) Development of large DNA methods for plants: molecular cloning of large segements of *Arabidopsis* and carrot DNA into yeast. *Nucleic Acids Research* **16**, 11091–11105.

Harrison. B.J. and Fincham, J.R.S. (1968) Instability at the *pal* locus in *Antirrhinum majus*. 3. A gene controlling mutation frequency. *Heredity* **23**, 67–77.

Hehl, R. and Baker, B. (1989) Induced transposition of Ds by a stable Ac in crosses of transgenic tobacco plants. *Molecular and General Genetics* **217**, 53–59.

Klee, H., Horsch, R. and Rogers, S. (1987) *Agrobacterium*-mediated plant transformation and its further applications to plant biology. *Annual Review of Plant Physiology* **38**, 467–486.

Knapp, S., Coupland, G., Uhrig, H., Starlinger, P. and Salamini, F. (1988) Transposition of the maize transposable element Ac in *Solanum tuberosum*. *Molecular and General Genetics* **213**, 285–290.

Knox, R.B. (1984) Pistil–pollen interactions. In *Cellular Interactions: Encyclopedia of Plant Physiology*, Vol. 17, ed. H.F. Linskens and J. Heslop-Harrison. Springer Verlag, Berlin, pp. 508–608.

Kobayashi, D.Y., Tamaki, S.J. and Keen, N.T. (1989) Cloned avirulence genes from the tomato pathogen *Pseudomonas syringae* pv. *tomato* confer cultivar specificity on soybean. *Proceedings of the National Academy of Sciences (USA)* **86**, 157–161.

Koncz, C., Martini, N., Mayerhofer, R., Koncz-Kalman, Z., Korber, H., Redei, G.P. and Schell, J. (1989) High-frequency T-DNA-mediated gene tagging in plants. *Proceedings of the National Academy of Sciences (USA)* **86**, 8467–8471.

Lamb, C.J., Lawton, M.A., Dron, M. and Dixon, R.A. (1989) Signals and transduction mechanisms for activation of plant defences against microbial attack. *Cell* **56**, 215–224.

McClintock, B. (1951) Chromosome organisation and gene expression. *Cold Spring Harbor Symposia on Quantitative Biology* **16**, 13–47.

McClintock, B. (1965) The control of gene action in maize. *Brookhaven Symposium Biology* **18**, 162–184.

Maize Genetic Co-operative Newsletter (1986) **60**, 40–45.

Marks, M.D. and Feldmann, K.A. (1989) Trichome development in *Arabidopsis thaliana*. I. T-DNA tagging of the *GLABROUS 1* gene. *Plant Cell* **1**, 1043–1050.

Martin, C., Carpenter, R., Sommer, H., Saedler, H. and Coen, E.S. (1985)

Molecular analysis of instability in flower pigmentation of *Antirrhinum majus* following isolation of the *pallida* locus by transposon tagging. *EMBO Journal* **4**, 1625–1630.

Martin, C., Prescott, A., Lister, C. and MacKay, S. (1989) Activity of the transposon Tam3 in *Antirrhinum* and tobacco: possible role of DNA methylation. *EMBO Journal* **8**, 997–1104.

Mayo, G.M.E. and Shepherd, K.W. (1980) Studies of genes controlling specific host–parasite interactions in flax and its rust. I. Fine structure of the M group in the host. *Heredity* **44**, 211–227.

Meyerowitz, E. and Pruitt, R.E. (1985) *Arabidopsis thaliana* and plant molecular genetics. *Science* **229**, 1214–1218.

Nevers, P., Shepherd, N.S. and Saedler, H. (1986) Plant transposable elements. *Advances in Botanical Research* **12**, 103–203.

Newbury, H.J. (1986) Multiplication of *Antirrhinum majus* L. by shoot tip culture. *Plant Cell, Tissue and Organ Culture* **7**, 39–42.

O'Reilly, C., Shepherd, N.S., Pereira, A., Schwartz-Sommer, Zs., Bertram, I., Robertson, D.S., Peterson, P.A. and Saedler, H. (1985) Molecular cloning of the A1 locus of *Zea mays* using the transposable elements En and Mu1. *EMBO Journal* **4**, 877–882.

Paz-Ares, J., Wienand, U., Peterson, P.A. and Saedler, H. (1986) Molecular cloning of the C locus of *Zea mays*: a locus regulating the anthocyanin pathway. *EMBO Journal* **5**, 829–833.

Pereira, A. and Saedler, H. (1989) Transpositional behaviour of the maize En-1/ Spm element in transgenic tobacco. *EMBO Journal* **8**, 1315–1321.

Pryor, A. (1987a) Stability of alleles of Rp (resistance to *Puccinia sorghi*). *Maize Genetic Co-operative Newsletter* **61**, 37–38.

Pryor, A. (1987b) The origin and structure of fungal resistance genes in plants. *Trends in Genetics* **3**, 157–161.

Sampson, D.R. (1960) Linkage of a rust resistance gene with two flower color genes in *Antirrhinum majus*. *Canadian Journal of Genetic Cytology* **2**, 216–219.

Saxena, K.M. and Hooker, A.L. (1968) On the structure of a gene for disease resistance in maize. *Proceedings of the National Academy of Sciences* **61**, 1300–1305.

Schmidt, R. and Willmitzer, L. (1989) The maize autonomous element Activator (Ac) shows a minimal germinal excision frequency of 0.2%–0.5% in transgenic *Arabidopsis thaliana* plants. *Molecular and General Genetics* **220**, 17–24.

Schmidt, R.J., Burr, F.A. and Burr, B. (1987) Transposon tagging and molecular analysis of the maize regulatory locus opaque-2. *Science* **238**, 960–963.

Shepherd, N.S. (1988) Transposable elements and gene-tagging. In *Plant Molecular Biology – a Practical Approach*, ed. C.H. Shaw. IRL Press, Oxford. ch. 8, pp. 189–220.

Staskawicz, B.J., Dahlbeck, D. and Keen, N.T. (1984) Cloned avirulence gene of *Pseudomonas syringae* pv *glycinea* determines race-specific incompatibility in *Glycine max* (L.) Merr. *Proceedings of the National Academy of Sciences (USA)* **81**, 6024–6028.

Staskawicz, B., Dahlbeck, D., Keen, N. and Napoli, C. (1987) Molecular characterisation of cloned avirulence genes from race 0 and race 1 of *Pseudomonas syringae* pv. *glycinea*. *Journal of Bacteriology* **169**, 5789–5794.

Steinmentz, M., Minard, K., Harvarth, S., McNicholas, J., Srelinger, J., Wake, C., Long, E., Mach, B. and Hood L. (1982) A molecular map of the immune response region from the major histoincompatibility complex of the mouse. *Nature* **300**, 35–42.

Swanson, J., Kearney, B., Dahlbeck, D. and Staskawicz, B. (1988) Cloned avirulence gene of *Xanthomonas campestris* pv. *vesicatoria* complements spontaneous race-change mutants. *Molecular Plant–Microbe Interactions* **1**, 5–9.

Theres, N.W. (1986) Inaugural dissertation, University of Cologne.

Van Sluys, M.A., Tempe, J. and Federoff, N. (1987) Studies on the introduction and mobility of the maize *Activator* element in *Arabidopsis thaliana* and *Daucus carota. EMBO Journal* **6**, 3881–3889.

Wienand, U., Weydemann, U., Niesbach-Kloesgen, U., Peterson, P.A. and Saedler, H. (1986) Molecular cloning of the *c2* locus of *Zea mays*, the gene coding for chalcone synthase. *Molecular and General Genetics* **203**, 202–207.

Wiese, L. (1984) Mating systems in unicellular algae. In *Cellular Interactions: Encyclopedia of Plant Physiology*, Vol. 17, ed. H.F. Linskens and J. Heslop-Harrison. Springer Verlag, Berlin, pp. 239–260.

Yoder, J.I., Palys, J., Alpert, K. and Lassner, M. (1988) Ac transposition in transgenic tomato plants. *Molecular and General Genetics* **213**, 291–296.

Young, N.D., Zamir, D., Ganal, M.W. and Tanksley, S.D. (1988) Use of isogenic lines and simultaneous probing to identify DNA markers tightly linked to the *Tm-2a* gene in tomato. *Genetics* **120**, 579–585.

Chapter 6
Engineering of Insect-resistant Plants with *Bacillus thuringiensis* Crystal Protein Genes

Marnix Peferoen

Plant Genetic Systems N.V., J. Plateaustraat 22,
9000 Gent, Belgium

1. *Bacillus thuringiensis,* the insecticidal bacterium

1.1 The history

The first record on *Bacillus thuringiensis* goes back to 1901, when Ishiwata discovered a bacterium from diseased silkworm larvae. Unfortunately, the original strain was lost and is now referred to as *Bacillus sotto*. In 1909, Dr Berliner received some diseased Mediterranean flour moth larvae from a mill in Thüringen. From these larvae he was able to isolate a sporulating bacterium and he named the bacterium *Bacillus thuringiensis*. He also noticed that the bacterium had a specific activity since it was toxic to Mediterranean flour moth (*Ephestia kühniella*) larvae and not to mealworm (*Tenebrio molitor*) larvae (Berliner, 1915). After World War I, some work was done on the use of *Bacillus thuringiensis*, primarily against the European corn borer in South-East Europe. After World War II, as well as in Europe, *Bacillus thuringiensis* was also evaluated for its potency in the control of several insect pests in the USSR and North America. In 1960, stimulated by the growing concern over the use of chemical insecticides, the first *Bacillus thuringiensis* strain was commercialized, and was marketed as 'Thuricide'. This strain was later gradually replaced with a more potent strain (HD-1) isolated by Dulmage (1970), and the application was extended to many different lepidopteran pests. It was generally accepted that *Bacillus thuringiensis* was only active against Lepidoptera, until Goldberg and Margalit isolated a strain (*Bacillus thuringiensis israelensis*) from a mosquito breeding pond in the Negev desert, which proved highly toxic to mosquito and black fly larvae (Goldberg and

135

Margalit, 1977). In 1983, Krieg and co-workers isolated a strain from a dead mealworm pupa (*Bacillus thuringiensis tenebrionis*) which was highly toxic to elm leaf beetle and Colorado potato beetle larvae. Both findings were a clear stimulus, screening programmes were started and there are now thousands of *Bacillus thuringiensis* isolates in various collections.

1.2 The classification

As early as the 1950s, it became clear that there was a need to design identification methods in order to classify the numerous strains isolated. Based on the agglutination reactions of motile *B. thuringiensis* cells with antisera generated against flagellae on vegetative cells, de Barjac and Bonnefoi (1962) designed the serotyping method. Some 30 different serotypes have been described, and the number is still growing (de Barjac and Franchon, 1990). Despite the fact that the biological significance of the serotype is unclear, the method is still used for general classification of *B. thuringiensis* strains. However, the serotype has certainly very little to do with the insecticidal activity of the strain. The *morrisoni* serotype, for example, contains strains active against lepidopteran, dipteran and coleopteran larvae (HD-12, PG-14 and *B. thuringiensis tenebrionis*, respectively).

2. *Bacillus thuringiensis*, the crystal proteins

2.1 A family of proteins

Berliner already noticed the occurrence of small inclusion bodies next to the spore (Berliner, 1915). In fact, the production of crystals upon sporulation is a general characteristic of all *B. thuringiensis* strains. However, only in 1953 was it demonstrated that the crystals are responsible for the insecticidal activity of *Bacillus thuringiensis* (Hannay, 1953). Soon thereafter, Hannay and Fitz-James (1955) showed that the toxicity of the crystals was determined by alkaline soluble proteins. The final proof of the insecticidal activity of the crystal proteins was provided by Schnepf and Whiteley (1981). They cloned a crystal protein gene from *B. thuringiensis* in *E. coli*, producing a protein highly toxic to tobacco hornworm larvae.

Today 45 nucleotide sequences of crystal protein genes have been determined. Although they clearly constitute a family of related genes, except for the *cytA* gene, which is rather a cytotoxin than an insecticidal protein, there are 17 distinctly different crystal protein genes (Table 6.1). The *cry* genes encode proteins, either of some 130 kDa or some 70 kDa, which are proteolytically activated to a toxic core fragment of some

Table 6.1. Insecticidal crystal protein genes of *Bacillus thuringiensis.*

cry	Protein (mol. mass)	*B. thuringiensis* strain	Reference
crylA(a)	133.2	HD-1	Schnepf *et al.* (1985)
crylA(b)	131.0	Berliner 1715	Wabiko *et al.* (1986)
crylA(c)	133.3	HD-73	Adang *et al.* (1985)
crylB	138.0	HD-2	Brizzard and Whiteley (1988)
crylC	134.8	HD-110	Honée *et al.* (1988)
crylD	132.5	HD-68	Höfte *et al.* (1990)
crylE	132.5	HD-146	Plant Genetic Systems, unpublished
cryllA	70.9	HD-263	Widner and Whiteley (1989)
cryllB	70.8	HD-1	Widner and Whiteley (1989)
crylllA	73.1	*tenebrionis*	Herrnstadt *et al.* (1987)
crylllB	74.2	*tolworthi*	Sick *et al.* (1990)
crylllC	129.4	*galleriae*	Plant Genetic Systems, unpublished
crylVA	134.4	*israelensis*	Ward and Ellar (1987)
crylVB	127.8	*israelensis*	Chungjatupornchai *et al.* (1988)
crylVC	77.8	*israelensis*	Thorne *et al.* (1986)
crylVD	72.4	*israelensis*	Donovan *et al.* (1988)
cytA	27.4	*isrealensis*	Waalwijck *et al.* (1985)

60 kDa, with the exception of CryIVD (Chilcott and Ellar, 1988). The toxic fragment of the 130 kDa proteins is localized in the N-terminal half, so that apparently the highly conserved C-terminal half is not essential for toxicity. It is speculated that the C-terminal half, containing several cysteine residues, may be involved in the crystal formation. The proteins of some 70 kDa lack the C-terminal half of the 130 kDa proteins, and therefore appear as naturally truncated. None of these truncated proteins crystallize in the typical bipyramidal shape of most of the 130 kDa proteins. Through deletion analysis, the smallest toxic fragment of CryIA(b) was determined (Höfte *et al.*, 1986) and its 12 C-terminal amino acids proved to be highly conserved among Cry toxins (Schnepf and Whiteley, 1985; Pao-intara *et al.*, 1988). In addition to that region, there are four domains of highly conserved sequences in the toxic core fragments of all but three crystal proteins (CryIIA, CryIIB and CryIVD). The N-terminus is characterized by a hydrophobic region of some 120 amino acids in all crystal proteins, with the exception of CryII and CryIVD proteins. Based on the sequences of seven different toxins, a secondary structure model shows the N-terminus of the toxins to be rich in α-helix structures while the C-terminus contains alternating β-strands and coil structures, which are typical for a β-sheet conformation (Convents *et al.*, 1990).

Each crystal protein is characterized by its own particular insecticidal activity and spectrum. The crystal proteins are grouped in four classes, Lepidoptera-specific, Lepidoptera- and Diptera-specific, Coleoptera-specific and Diptera-specific. In the lepidopteran pathotype (CryI proteins), seven genes have been identified, coding for protoxins of 130 to 140 kDa. Three crystal proteins (CryIA(a), CryIA(b) and CryIA(c)) are very closely related (82 to 90% amino acid identity) and belong to a subfamily. All CryI proteins, even the CryIA subfamily, have a distinctive insecticidal spectrum (Höfte and Whiteley, 1989).

The CryII proteins, formerly designated as P2 proteins (Yamamoto and McLaughlin, 1981), are toxic to both lepidopteran and dipteran larvae (CryIIA) or to lepidopteran larvae (CryIIB) (Widner and Whiteley, 1989). These 65 kDa proteins accumulate in cuboidal crystals, and their homology to the other Cry proteins is rather limited.

The CryIIIA protein was the first crystal protein found to be toxic to coleopteran larvae such as the Colorado potato beetle (Herrnstadt *et al.*, 1987). Two translation initiation sites give rise to two proteins of 72 kDa and 66 kDa, which are both stored in rhomboid crystals (McPherson *et al.*, 1988). The CryIIIB protein, with 66% of its amino acids identical to CryIIIA, accumulates as a protein of some 74 kDa, in rhomboidal crystals and is also highly toxic to Colorado potato beetle larvae (Sick *et al.*, 1990). The CryIIIC protein, produced as a protoxin of some 129 kDa, crystallizes in bipyramidal crystals (Plant Genetic Systems (PGS), unpublished). The protoxin is proteolytically activated to a fragment of some 66 kDa, which is toxic to Colorado potato beetle larvae. The toxic fragment has some 37 and 32% homology to CryIIIA and CryIIIB, respectively. The C-terminal half of the protoxin is homologous to the corresponding part of other protoxins.

Two of the crystal proteins toxic to dipteran larvae (CryIV) are produced as protoxins of some 130 kDa (CryIVA and CryIVB) while the other two proteins (CryIVC and CryIVD) are naturally truncated (78 kDa and 72 kDa, respectively). Again, the protoxins are proteolytically activated to a toxic fragment of undetermined molecular weight (Chilcott and Ellar, 1988; Chungjatupornchai *et al.*, 1988).

Apart from these well-characterized crystal proteins, there are several proteins which do not seem to be active against any of the insects tested (Ohba *et al.*, 1987; Martin and Travers, 1989). However, it should be stressed that bioassays are focused on insects which are easy to test, are a threat to human health, or are economically important.

Most *cry* genes are situated on large conjugative plasmids (Lereclus *et al.*, 1989), and several of these genes have been shown to be bordered by transposon-like elements (Kronstad and Whiteley, 1984; Mahillon *et al.*, 1985; Bourgouin *et al.*, 1988). Analysis of the crystal protein composition of several lepidopteran strains demonstrated that many strains produce several crystal proteins and that the same crystal proteins can be found in strains of different serotypes (Höfte *et al.*, 1988a). They also reported a manifest correlation between the occurrence of a certain crystal protein and the insecticidal spectrum of the isolate, concluding that the insecticidal activity and spectrum of a strain are primarily determined by the crystal proteins.

2.2 The search for new crystal proteins

Today, the importance of *Bacillus thuringiensis* in an environmentally sound insect control programme is well accepted. Therefore, there is an active interest to extend the use of *Bacillus thuringiensis* strains. Inspired by the discovery of strains with activities against insects other than Lepidoptera, screening programmes for new *Bacillus thuringiensis* strains have been launched. Apart from trying to find new genes available in natural populations of *Bacillus thuringiensis*, research has been initiated to engineer synthetic crystal protein genes with improved insecticidal activity or insecticidal spectrum. So far, there are no reports on synthetic genes encoding crystal proteins with increased insecticidal activity or with a different insecticidal spectrum. Recently, experiments have been performed to delineate the regions on crystal proteins which are responsible for the toxicity and specificity (Ge *et al.*, 1989; Widner and Whiteley, 1990). Understanding the functional–structural relationships of crystal proteins is imperative for an intelligent engineering programme of crystal proteins.

3. Mechanism of action and insect resistance

3.1 A receptor-mediated specificity

The major steps involved in the mechanism of action of crystal proteins in lepidopteran larvae have been described by Lüthy and Ebersold (1981). Upon ingestion, the crystal proteins are dissolved by the alkaline gut juices in the midgut lumen, and are converted by gut proteases into toxic core fragments. Shortly thereafter, the midgut epithelial cells swell and eventually burst. In the last 5 years, the different steps, especially the interaction of crystal proteins with insect cells, have been studied in great detail.

Most studies indicate that the specificity of the crystal proteins is not determined by proteolytic activation in the midgut (Lüthy and Ebersold, 1981; Jaquet *et al.*, 1987). However, Haider *et al.* (1986) demonstrated that, depending on the proteolytic enzymes, a protoxin can be activated into either a dipteran or a lepidopteran toxin. Choma *et al.* (1990) reported that the toxic fragment of CryIA(c) is generated by different specific sequential proteolytic cleavages, a process which could be different in each insect species and which could in part explain differences in the host spectrum. Work by Arvidson *et al.* (1989) suggests that specificity is lost upon reduction of the cysteine residues in the protoxin, but can be restored by reoxidation of the cysteines and/or proteolytic removal of the C-terminus containing most cysteine residues.

Although differential processing may be involved, recent experiments show that the interaction of toxins with high-affinity binding sites in the brush border of midgut epithelial cells predominantly determines the specificity. Binding studies with brush border membrane vesicles, prepared from *Manduca sexta* and *Pieris brassicae* midgut epithelial cells, and labelled proteins demonstrated that there is a distinct correlation between the binding characteristics of a certain crystal protein and its toxicity (Hofmann *et al.*, 1988). Experiments with *Heliothis virescens* and *Spodoptera littoralis* confirmed that toxicity of a crystal protein correlates with a high-affinity binding to a significant number of binding sites (Van Rie *et al.*, 1990a, b). These findings are further corroborated by the results of binding studies with insects resistant to a certain crystal protein. In contrast, Wolfersberger (1990) found an inverse correlation between receptor affinity and toxicity in gypsy moth larvae. This finding suggests that the toxicity is only partially determined by the binding characteristics of the crystal proteins. Specific binding to midgut receptors may be essential for the crystal proteins to be toxic; the level of toxicity could be determined by the efficiency of the pore formation in the epithelial membrane.

It is not very clear what happens after the crystal protein has interacted with its receptor. Sacchi *et al.* (1986) studied the effect of toxins on the potassium–amino acid co-transport in brush border membrane vesicles prepared from midguts of *Pieris brassicae* and measured a dissipation of the potassium gradient by the formation of channels in the brush border membranes. They concluded that the channels were potassium-specific. However, Knowles and Ellar (1986) proposed the mechanism of colloid–osmotic lysis, which involves the formation of small (0.5–1 nm) non-specific pores in the cell membrane of the epithelial cells, resulting in an influx of ions accompanied by an influx of water. As a consequence, the cells swell and eventually lyse. The formation of non-specific pores, rather than potassium-selective channels, was confirmed by Hendrickx *et al.* (1990). So far, it is unclear whether *B. thuringiensis* toxins induce pore formation indirectly through interaction with membrane molecules, or directly through insertion into the membrane.

3.2 The development of resistance in insects

Resistance to some *Bacillus thuringiensis* crystal proteins has been induced in laboratory cultures of *Plodia interpunctella* (McGaughey, 1985), *Cadra cautella* (McGaughey and Beeman, 1988) and *Heliothis virescens* (Stone *et al.*, 1989). *Plodia interpunctella* and *Cadra cautella* were selected by continuous exposure to high levels of *Bacillus thuringiensis*, resulting in a > 250-fold increase of resistance in *P. interpunctella* after 36 generations, and a seven-fold increase of resistance in *C. cautella* after 21 generations (McGaughey and Beeman, 1988). In *Heliothis virescens*, a 24-fold increase in resistance was obtained after seven generations following continuous exposure to a crystal protein (CryIA(c)) at a concentration necessary to kill 66% of the population. In Hawaiian fields intensely treated for many years with a *Bacillus thuringiensis* spray, Tabashnik *et al.* (1990) found up to a 41-fold increase of resistance in *Plutella xylostella* populations. Obviously, the study of these resistant insects would not only yield valuable data on the mechanism of action of crystal proteins, but could also lead to strategies to retard the build-up of resistance to crystal proteins in insect populations.

Although the *Plodia interpunctella* larvae developed resistance to certain *B. thuringiensis* strains, they were still susceptible to some other strains (McGaughey and Johnson, 1987). Surprisingly, experiments with purified crystal proteins demonstrated that the *P. interpunctella* strain resistant to the CryIA(b) toxin had become more susceptible to the CryIC toxin (Van Rie *et al.*, 1990c). Resistance to the CryIA(b) protein correlates with a 50-fold reduction in affinity of the membrane receptor for this protein. Increased susceptibility to the CryIC toxin corresponds to an increase of the CryIC-binding sites in the midgut epithelial cells (Table 6.2). This clearly demonstrates that insects can develop resistance to a

Table 6.2. Correlation between binding characteristics and toxicity of *Bacillus thuringiensis* crystal proteins in *Plodia interpunctella*.

Cry	P. interpunctella-susceptible			P. interpunctella-resistant		
	LD_{50}	K_d	R_t	LD_{50}	K_d	R_t
CryIA(b)	0.03	0.72	1.44	> 12.8	36.3	1.77
CryIC	0.11	0.31	0.38	0.03	0.18	1.15

LD_{50}: the dose required to kill 50% of the insects tested (µg protein per larva).
K_d: binding site dissociation constant (nmol).
R_t: binding site concentration (pmol per mg membrane protein).

crystal protein by changing the binding characteristics of the midgut receptors. These experiments have also shown that resistance to one crystal protein does not necessarily cause resistance to other crystal proteins. *Bacillus thuringiensis* crystal proteins are a family of related proteins, possibly mirrored by a family of related receptors in the insect midgut.

4. Engineering of insect-resistant plants

4.1 Plant engineering

Since the first transgenic tobacco plants expressing foreign proteins were obtained (Herrera-Estrella *et al.*, 1984; Horsch *et al.*, 1985), there has been significant progress in tissue culture, plant transformation and plant molecular biology. Most engineered plants have been generated through the *Agrobacterium tumefaciens*-mediated gene transfer, because it is a very efficient and versatile vector for the stable introduction of genes into plants (Hooykaas, 1989). So far *Agrobacterium tumefaciens*, a plant pathogen casing tumorous crowngalls on infected dicotyledonous plants, seems unsuccessful in the transformation of most monocotyledonous plants, especially of cereals. For those plants resistant to *Agrobacterium*-mediated transformation, alternative DNA delivery systems, such as polyethylene glycol-mediated transfer (Uchimiya *et al.*, 1986), microinjection (De la Pena *et al.*, 1987), electroporation (Toriyama *et al.*, 1988) and particle gun technology (Klein *et al.*, 1988), are being contemplated. The particle gun technology looks especially promising, primarily because it allows the direct delivery of DNA into a wide range of plant cells, obviating the regeneration of plants from transformed protoplasts (Potrykus, 1990). Today, some 25 different plant species have been transformed using different systems (Gasser and Fraley, 1989; Uchimiya *et al.*, 1989).

In the past 10 years, plant molecular research has led to a better understanding of gene regulation. Foreign genes can only be expressed in plants when the gene is surrounded by appropriate signals. Several promoters have been described (Herrera-Estrella *et al.*, 1984; Velten *et al.*, 1984; Odell *et al.*, 1985; De Almeida *et al.*, 1989), some of which confer expression in a wide variety of tissues during most stages of development, others being very specific for a particular tissue, a certain stimulus and/or a developmental stage (Benfey and Chua, 1989). At the same time, considerable progress has also been made in the identification of genes to be used in the engineering of plants for crop improvements, such as herbicide resistance (Botterman and Leemans 1988), virus resistance (Lawson *et al.*, 1990) and of course insect resistance (Hilder *et al.*, 1987; Vaeck *et al.*, 1987; Johnson *et al.*, 1989).

4.2 Expression of crystal proteins

The TR promoter from the Ti plasmid of *Agrobacterium tumefaciens* (Velten *et al.*, 1984) and the cauliflower mosaic virus (CaMV)35S promoter (Odell *et al.*, 1985) are two promoters which have been used in the transformation of plants with crystal proteins. The TR promoter, a dual promoter directing expression in both directions, normally confers very low levels of expression in plants. However, upon wounding, transcription is stimulated, resulting in high levels of expression (Teeri *et al.*, 1989). The 35S promoter is generally considered to be a constitutive promoter, active during most stages of development and in most plant tissues. However, domains have been identified within this promoter which can confer different developmental and tissue-specific expression patterns (Benfey *et al.*, 1989).

In plant transformation, the *neo* gene encoding neomycin phospho-transferase II (NPTII), conferring resistance to neomycin/kanamycin, is commonly used as a selectable marker. The *neo* gene not only is used to select transformed plants, but can also be used to single out plants with high levels of expression. Using the dual TR promoter, the *neo* gene and the toxin gene can be cloned on both sides of the TR promoter. Since transcription is simultaneously initiated in both directions, selection for high kanamycin resistance also selects plants with high levels of crystal proteins. Furthermore, it proved possible to make a translational fusion between the crystal protein gene and the *neo* gene, resulting in a fusion protein toxic to insects and with NPTII activity (Höfte *et al.*, 1988b). A high level of kanamycin resistance should then evidently correspond to a high level of crystal protein.

Reports have been published on the engineering of tobacco, tomato and potato plants with four different crystal protein genes. The CryIA(b) crystal protein was expressed under the control of the TR promotor in tobacco (Vaeck *et al.*, 1987) and in potato (Peferoen *et al.*, 1990). The 35S promotor was used for the expression of CryIA(a) in tobacco (Barton *et al.*, 1987) and for the expression of CryIA(c) in tomato (Fischhoff *et al.*, 1987). McPherson *et al.*(1989) reported the expression of the CryIIIA crystal protein in potato. Compared with other genes transferred to plants, crystal protein genes are weakly expressed in transgenic plants. The tobacco, tomato and potato plants, transformed with the full-length sequences encoding the protoxin under the control of either the 35S or the TR promoter, have extremely low levels of crystal proteins. It appears that the crystal protein-encoding region itself reduces the levels of expression. Since only part of the crystal protein is necessary for the toxicity, plants can be engineered with truncated sequences, encoding the toxic fragment only. Expression levels are significantly enhanced by transforming with the truncated sequences. With truncated genes, up to 42 ng CryIA(b) per mg total leaf protein was measured, while with full length genes some 1 to

2 ng could be detected (Vaeck *et al.*, 1987). It is not fully clear why crystal protein genes, even truncated sequences, are expressed at such low levels. However, in contrast to most plant genes, crystal protein genes are extremely rich in AT nucleotides. It is therefore speculated that the codon usage in the crystal protein genes is far from optimal for plant expression, that there are polyadenylation signals within the coding region, and that the stability of the transcript is limited. Experiments to increase expression levels of crystal proteins in plants have been initiated, but no details of these experiments have been published as yet.

5. Insect-resistant plants

5.1 Laboratory evaluation

The first results on insect control by plants engineered with crystal protein genes were published in 1987 (Barton *et al.*, 1987; Fischhoff *et al.*, 1987; Vaeck *et al.*, 1987). The insecticidal activity of the tobacco plants was tested with tobacco hornworm (*Manduca sexta*) larvae, together with tobacco budworm (*Heliothis virescens*), its major pest. Vaeck *et al.* (1987) reported a clear correlation between the level of expression of the crystal proteins and the insecticidal activity. Tobacco plants expressing crystal protein as some 0.004% of their total leaf protein killed all larvae within 6 days. Engineered tomato plants were tested by exposure to tobacco hornworm (*Manduca sexta*) and tomato fruitworm (*Heliothis virescens*) larvae, and proved well protected against attacks from both insects (Fischhoff *et al.*, 1987). Similar results were also obtained by PGS as illustrated in Figure 6.1. Potatoes engineered with *cryIIIA* showed significant levels of resistance to feeding damage by Colorado potato beetle larvae (*Leptinotarsa decemlineata*) (McPherson *et al.*, 1989). In North Africa and Central to South America, potato tuber moth (*Phthorimaea operculella*) larvae are a serious pest on both potato plants and tubers. Leaves and tubers from different potato cultivars, engineered with the *cryIA(b)* gene, proved highly resistant to feeding and tunnelling damage by tuber moth larvae (Peferoen *et al.*, 1990 (Figure 6.2)).

5.2 Field evaluations

The first field trial with insect-resistant plants was done in 1986 in North Carolina with tobacco plants engineered by Plant Genetic Systems. Since then, tobacco plants and their progeny have been tested in the field for their resistance against tobacco hornworm and tobacco budworm damage. Plants with a moderate insecticidal activity in the greenhouse still proved

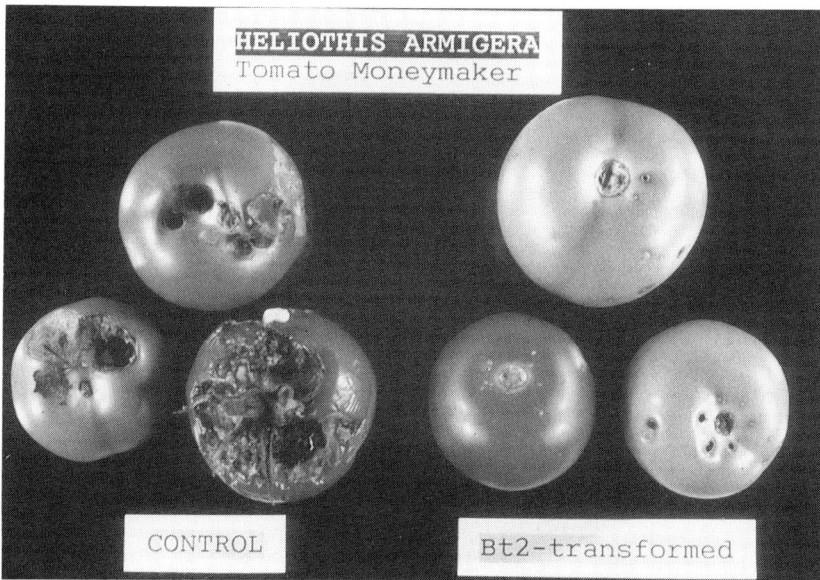

Figure 6.1. Damage to tomato fruits of Moneymaker (control) and Cry1A(b) engineered Moneymaker by *Heliothis armigera* larvae.

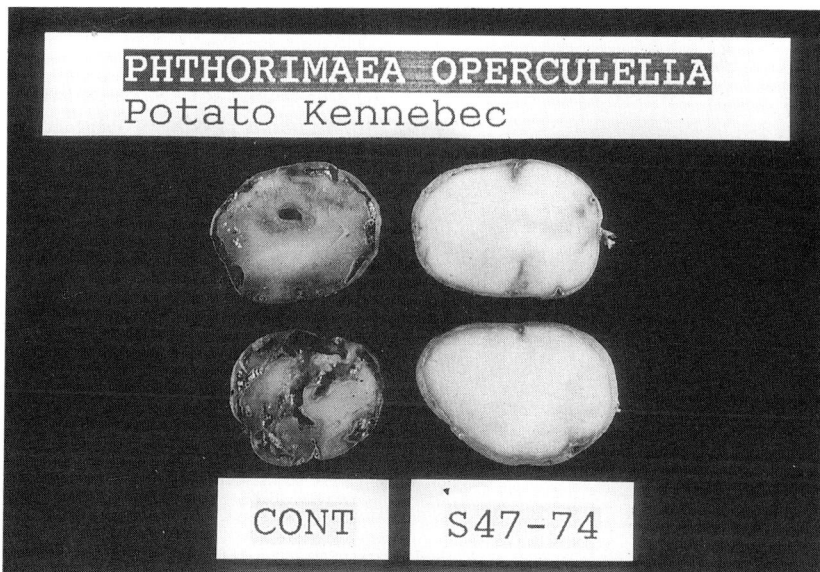

Figure 6.2. Damage to Kennebec (Cont) and Cry1A(b) engineered Kennebec potato tubers (S47-74) by potato tuber moth larvae (*Phthorimaea operculella*).

highly resistant to insect attacks in the field. Surviving larvae on these plants were heavily infested by parasitic wasps. The higher levels of pest control in the field, compared with results obtained under contained environmental conditions, could be the consequence of a synergistic effect between the toxin expressed by the plant and the presence of lepidopteran predators and parasites.

Delannay *et al.* (1989) tested tomato plants engineered with crystal proteins in the field for their resistance to important lepidopteran pests on tomato. Tomato plants can withstand some feeding damage to their leaves without significant yield losses, but the tomato fruits, especially fresh market tomatoes, should be virtually undamaged. Tomato fruitworm and tomato pinworm larvae may cause some damage to leaves but primarily bore through the tomato skin and enter the fruit. Tomato fruits of the engineered plants suffered less fruit damage, even under heavy infestation conditions of fruitworm and pinworm larvae. However, the damage to the fruits was still significant and commercially unacceptable. Again, engineered plants seemed to perform better in the field than in the greenhouse, corroborating the indication of a synergy between crystal proteins in the plants and natural predators and parasites.

Cotton transformed with crystal protein genes has also been tested in the field, but no data have been published as yet.

6. Perspectives

Since the first reports, the potentials and the limitations of engineering insect resistance in plants with *Bacillus thuringiensis* crystal protein genes have been intensely explored. Crops such as tobacco, tomato and potato can be protected against insects which are highly susceptible to *Bacillus thuringiensis* crystal proteins. It is expected that these crops, with an engineered protection against some of their major pest insects, will be the first commercial products to be marketed in the mid-1990s. These pioneering products will have to pave the way for registration and to attain acceptance by the breeders, the farmers and the consumers.

In order to ensure a prolonged effectiveness, and to extend the application to a wider range of crops and a wider range of insects, there is need for progress in different areas. Several scientists have raised concerns about the potential for the development of resistance to *Bacillus thuringiensis* crystal proteins in genetically engineered plants (Gould, 1988). The fact that different crystal proteins can bind to different receptors in the insect midgut suggests that a combination or alternation of different proteins could help to retard and possibly to prevent the development of resistance. Selection pressure on insect populations could also be reduced by limiting the expression of crystal proteins to the economically important organs of the plants, allowing limited damage to less important parts. In the coming years, research will be focused on the crystal protein receptors and on the genetics of the resistance. This information will be essential in order to design strategies that will counteract resistance development in insects (Tabashnik, 1989). Several important crops, especially the monocotyledonous plants, are still very difficult to transform. Recent successes, such as soyabean (McCabe *et al.*, 1988) and rice transformation (Shimamoto *et al.*, 1989), indicate that transformation of several other crops could be achieved in the near future. Several important pest insects are not susceptible to currently available *Bacillus thuringiensis* crystal proteins. Screening programmes could, as they did in the past, lead to the discovery of crystal proteins toxic to some of these insects. Obviously, *Bacillus thuringiensis* will not be the answer to all the insect problems. For some insects alternative insecticidal proteins will be needed, and are being actively pursued. Inhibitors of digestive enzymes (Hilder *et al.*, 1987; Altabella and Crispeels, 1990) seem to have some interesting potential. Because of the expression problems with crystal proteins in plants, moderately susceptible insects cannot be controlled effectively. In the last few years, research has been focused on this problem and we are on the verge of some major breakthroughs in this area. This on its own will considerably expand the use of crystal proteins in the engineering of insect resistance in plants.

References

Adang, L.F., Staver, M.J., Rocheleau, T.A., Leighton, J., Barker, R.F. and Thompson, D.V. (1985) Characterized full-length and truncated plasmid clones of the crystal protein of *Bacillus thuringiensis* subsp. *kurstaki* HD-73 and their toxicity to *Manduca sexta. Gene* **36**, 289–300.

Altabella, T. and Crispeels, M.J. (1990) Tobacco plants transformed with the bean *aai* gene express an inhibitor of insect α-amylase in their seeds. *Plant Physiology* **93**, 805–810.

Arvidson, H., Dunn, P.E., Strnad, S. and Aronson, A.I. (1989) Specificity of *Bacillus thuringiensis* for lepidopteran larvae: factors involved *in vivo* and in the structure of a purified protoxin. *Molecular Microbiology* **3**, 1533–1543.

Barton, K., Whiteley, H. and Yang, N.-S. (1987) *Bacillus thuringiensis* δ-endotoxin in transgenic *Nicotiana tabacum* provides resistance to lepidopteran insects. *Plant Physiology* **85**, 1103–1109.

Benfey, P.N. and Chua, N.-H. (1989) Regulated genes in transgenic plants. *Science* **244**, 174–181.

Benfey, P.N., Ren, L. and Chua, N.-H. (1989) The CaMV 35S enhancer contains at least two domains which can confer different developmental and tissue-specific expression patterns. *EMBO Journal* **8**, 2195–2202.

Berliner, E. (1915) Uber die Schlaffsucht der Mehlmottenraupe (*Ephestia kuehniella* Zell), und ihren Erreger *Bacillus thuringiensis* n. sp. *Zeitschrift für angewandtes Entomologie* **2**, 29–56.

Botterman, J. and Leemans, J. (1988) Engineering of herbicide resistance. *Trends in Genetics* **4**, 219–222.

Bourgouin, C., Delecluse, A., Ribier, J., Klier, A. and Rapoport, G. (1988) A *Bacillus thuringiensis* subsp. *israelensis* gene encoding a 125-kilodalton larvicidal polypeptide is associated with inverted repeat sequences. *Journal of Bacteriology* **170**, 3575–3583.

Brizzard, B.L. and Whiteley, H.R. (1988) Nucleotide sequence of an additional crystal protein gene cloned from *Bacillus thuringiensis* subsp. *thuringiensis*. *Nucleic Acids Research* **16**, 4168–4169.

Chilcott, C.N. and Ellar, D.J. (1988) Comparative study of *Bacillus thuringiensis* var. *israelensis* crystal proteins *in vivo* and *in vitro*. *Journal of General Microbiology* **134**, 2551–2558.

Choma, C.T., Surewicz, W.K., Carey, P.R., Pozsgay, M., Raynor, T. and Kaplan, H. (1990) Unusual proteolysis of the protoxin and toxin of *Bacillus thuringiensis* – structural implications. *European Journal of Biochemistry* **189**, 523–527.

Chungjatupornchai, W., Höfte, H., Seurinck, J., Angsuthanasombat, C. and Vaeck, M. (1988) Common features of *Bacillus thuringiensis* toxins specific for Diptera and Lepidoptera. *European Journal of Biological Chemistry* **173**, 9–16.

Convents, D., Houssier, C., Lasters, I. and Lauwereys, M. (1990) The *Bacillus thuringiensis* delta-endotoxin – evidence for a two domain structure of the minimal toxic fragment. *Journal of Biological Chemistry* **265**, 1369–1375.

De Almeida, E.R.P., Gossele, V., Muller, C.G., Dockx, J., Reynaerts, A., Botterman, J., Krebbers, E. and Timko, M.P. (1989) Transgenic expression of two marker genes under the control of an *Arabidopsis rbcS* promoter: sequences encoding the Rubisco transit peptide increase expression levels. *Molecular and General Genetics* **218**, 78–86.

de Barjac, H. and Bonnefoi, A. (1962) Essai de classification biochemique et sérologique de 24 souches de *Bacillus* du type *B. thuringiensis*. *Entomophaga* **1**, 5–31.

de Barjac, H. and Franchon, E. (1990) Classification of *Bacillus thuringiensis* strains. *Entomophaga* **35**, 233–240.

Delannay, X., LaVallee, B.J., Proksch, R.K., Fuchs, R.L., Sims, S.R., Greenplate, J.T., Marrone, P.G., Dodson, R.B., Augustine, J.J., Layton, J.G. and Fischhoff, D.A. (1989) Field performance of transgenic tomato plants expressing the *Bacillus thuringiensis* var. *kurstaki* insect control protein. *Bio/Technology* **7**, 1265–1269.

De la Pena, A., Lorz, H. and Schell, J. (1987) Transgenic rye plants obtained by injecting DNA into young floral tillers. *Nature* **325**, 274–276.

Donovan, W.P., Dankocsik, C.C. and Gilbert, M.P. (1988) Molecular characterization of a gene encoding a 72-kilodalton mosquito-toxic crystal protein from *Bacillus thuringiensis* subsp. *israelensis*. *Journal of Bacteriology* **170**, 4732–4738.

Dulmage, H.T. (1970) Insecticidal activity of HD-1, a new isolate of *Bacillus thuringiensis* var. *alesti*. *Journal of Invertebrate Pathology* **15**, 232–239.

Fischhoff, D.A., Bowdish, K.S., Perlak, F.J., Marrone, P.G., McCormick, S.M.,

Niedermeyer, J.G., Dean, D.A., Kusano-Kretzmer, K., Mayer, E.J., Rochester, D.E., Rogers, S.G. and Fraley, R.T. (1987) Insect tolerant transgenic tomato plants. *Bio/Technology* **5**, 807–813.

Gasser, C.S. and Fraley, R. (1989) Genetically engineering plants for crop improvement. *Science* **244**, 1293–1299.

Ge, A.Z., Shivarova, N.I. and Dean, D.H. (1989) Location of the *Bomyx mori* specificity domain on a *Bacillus thuringiensis* δ-endotoxin protein. *Proceedings of the National Academy of Sciences (USA)* **86**, 4037–4041.

Goldberg, L.J. and Margalit, J. (1977) A bacterial spore demonstrating rapid larvicidal activity against *Anopheles serengetii, Uranotaenia unguiculata, Culex univittatus, Aedes aegypti* and *Culex pipiens. Mosquito News* **37**, 355–358.

Gould, F. (1988) Evolutionary biology and genetically engineered crops. *Bioscience* **38**, 26–33.

Haider, M.Z., Knowles, B. and Ellar, D.J. (1986) Specificity of *Bacillus thuringiensis* var. *colmeri* insecticidal delta-endotoxin by differential processing of the protoxin by larval gut proteases. *European Journal of Biochemistry* **156**, 531–540.

Hannay, C.L. (1953) Crystalline inclusions in aerobic spore-forming bacteria. *Nature* **172**, 1004.

Hannay, C.L. and Fitz-James, P. (1955) The protein crystals of *Bacillus thuringiensis* Berliner. *Canadian Journal of Microbiology* **1**, 674–710.

Hendrickx, K., De Loof, A. and Van Mellaert, H. (1990) Effects of *Bacillus thuringiensis* delta-endotoxin on the permeability of brush border membrane vesicles from tobacco hornworm (*Manduca sexta*) midgut. *Comparative Biochemistry and Physiology* **95C**, 241–245.

Herrera-Estrella, L., Van Den Broeck, G., Maenhaut, R., Van Montagu, M., Schell, J., Timko, M. and Cashmore, A. (1984) Light-inducible and chloroplast-associated expression of a chimeric gene introduced into *Nicotiana tabacum* using a Ti plasmid vector. *Nature* **310**, 115–120.

Herrnstadt, C., Gilroy, T.E., Sobieski, D.A., Bennet, B.D. and Gaertner, F.H. (1987) Nucleotide sequence and deduced amino acid sequence of a coleopteran-active delta-endotoxin gene from *Bacillus thuringiensis* subsp. *san diego. Gene* **57**, 37–46.

Hilder, V.A., Gatehouse, A.M.R., Sheerman, S.E., Barker, R.F. and Boulter, D. (1987) A novel mechanism of insect resistance engineered in tobacco. *Nature* **300**, 160–163.

Hofmann, C., Vanderbruggen, H., Höfte, H., Van Rie, J., Jansens, S. and Van Mellaert, H. (1988) Specificity of *Bacillus thuringiensis* δ-endotoxins is correlated with the presence of high-affinity binding sites in the brush border membrane of target insect midguts. *Proceedings of the National Academy of Sciences (USA)* **85**, 7844–7848.

Höfte, H. and Whiteley, H.R. (1989) Insecticidal crystal proteins of *Bacillus thuringiensis. Microbiological Reviews* **53**, 242–255.

Höfte, H., De Greve, H., Seurinck, J., Jansens, S., Mahillon, J., Ampe, C., Vandekerckhove, J., Vanderbruggen, H., Van Montagu, M., Zabeau, M. and Vaeck, M. (1986) Structural and functional analysis of a cloned delta endotoxin of *Bacillus thuringiensis* Berliner 1715. *European Journal of Biochemistry* **161**, 273–280.

Höfte, H., Van Rie, J., Jansens, S., Van Houtven, A., Vanderbruggen, H. and Vaeck, M. (1988a) Monoclonal antibody analysis and insecticidal spectrum of three types of lepidopteran-specific insecticidal crystal proteins of *Bacillus thuringiensis*. *Applied and Environmental Microbiology* **54**, 2010–2017.

Höfte, H., Buyssens, S., Vaeck, M. and Leemans, J. (1988b) Fusion proteins wtih both insecticidal and neomycin phosphotransferase II activity. *FEBS Letters* **226**, 364–370.

Höfte, H., Soetaert, P., Jansens, S. and Peferoen, M. (1990) Nucleotide sequence and deduced amino acid sequence of a new Lepidoptera-specific crystal protein gene from *Bacillus thuringiensis*. *Nucleic Acids Research* **18**, 5545.

Honée, G., Salm, van der, T. and Visser, B. (1988) Nucleotide sequence of crystal protein gene isolated from *B. thuringiensis* subspecies entomocidus 60.5 coding for a toxin highly active against *Spodoptera* species. *Nucleic Acids Research* **16**, 6240.

Hooykaas, P.J.J. (1989) Transformation of plant cells via *Agrobacterium. Plant Molecular Biology* **13**, 327–336.

Horsch, R.B., Fry, J.E., Hoffman, N.L., Wallroth, M., Eichholtz, D., Rogers, S.G. and Fraley, R.T. (1985) A simple and general method for transferring genes into plants. *Science* **227**, 1229–1231.

Jaquet, F., Hütter, R. and Lüthy, P. (1987) Specificity of *Bacillus thuringiensis* delta-endotoxin. *Applied and Environmental Microbiology* **53**, 500–504.

Johnson, R., Narvaez, J., An, G. and Ryan, C. (1989) Expression of proteinase inhibitors I and II in transgenic tobacco plants: effects on natural defense against *Manduca sexta* larvae. *Proceedings of the National Academy of Sciences (USA)* **86**, 9871–9875.

Klein, T.M., Harper, E.C., Svab, Z., Sanford, J.C., Fromm, M.E. and Maliga, P. (1988) Stable genetic transformation of intact *Nicotiana* cells by the particle bombardment process. *Proceedings of the National Academy of Sciences (USA)* **85**, 8502–8505.

Knowles, B.H. and Ellar, D.J. (1986) Colloid–osmotic lysis is a general feature of the mechanism of action of *Bacillus thuringiensis* δ-endotoxins with different insect specificities. *Biochimica et Biophysica Acta* **924**, 509–518.

Krieg, A., Huger, A.M., Langenbruch, G.A. and Schnetter, W. (1983) *Bacillus thuringiensis* var. *tenebrionis*: ein neuer gegenüber Larven von Coleopteren wirksamer Pathotyp. *Journal of Applied Entomology* **96**, 500–508.

Kronstad, J.W. and Whiteley, H.R. (1984) Inverted repeat sequences flank the *Bacillus thuringiensis* crystal protein gene. *Journal of Bacteriology* **160**, 95–102.

Lawson, C., Kaniewski, W., Haley, L., Rozman, R., Newell, C., Sandes, P. and Tumer, N.E. (1990) Engineering resistance to mixed virus infection in a commercial potato cultivar; resistance to potato virus X and potato virus Y in transgenic Russet Burbank. *Bio/Technology* **8**, 127–134.

Lereclus, D., Bourgouin, C., Lecadet, M.M., Klier, A. and Rapoport, G. (1989) Role, structure, and molecular organization of the genes coding for the parasporal delta-endotoxins of *Bacillus thuringiensis*. In *Regulation of Procaryotic Development*, ed. I. Smith, R.A. Slepecky and P. Setlow. American Society for Microbiology, Washington DC, pp. 255–276.

Lüthy, P. and Ebersold, H.R. (1981) *Bacillus thuringiensis* delta-endotoxin:

histopathology and molecular mode of action. In *Pathogenesis of Invertebrate Microbial Diseases*, ed. E.W. Davidson. Allenheld, Osmun, and Co., Totawa, NJ. pp. 235–267.

McCabe, D.E., Swain, W.F., Martinell, B.J. and Christou, P. (1988) Stable transformation of soybean (*Glycine max*) by particle acceleration. *Bio/Technology* 6, 923–926.

McGaughey, W.H. (1985) Insect resistance to the biological insecticide *Bacillus thuringiensis*. *Science* 229, 193–195.

McGaughey, W.H. and Beeman, R.W. (1988) Resistance to *Bacillus thuringiensis* in colonies of Indeanmeal moth and almond moth (Lepidoptera: Pyralidae). *Journal of Economic Entomology* 81, 28–33.

McGaughey, W.H. and Johnson, D.E. (1987) Toxicity of different serotypes and toxins of *Bacillus thuringiensis* to resistant and susceptible Indeanmeal moths (Lepidoptera: Pyralidae). *Journal of Economic Entomology* 80, 1122–1126.

McPherson, S.A., Perlak, F.J., Fuchs, R.L., Marrone, P.G., Lavrik, P.B. and Fischhoff, D.A. (1988) Characterization of the coleopteran-specific protein gene of *Bacillus thuringiensis* var. *tenebrionis*. *Bio/Technology* 6, 61–66.

McPherson, S., Perlak, F., Fuchs, R., MacIntosh, S., Dean D. and Fischhoff, D. (1989) Expression and analysis of the insect control protein from *Bacillus thuringiensis* var. *tenebrionis*. In *Abstracts of the First International Symposium on the Molecular Biology of the Potato*, Bar Harbor, Maine, p. 51.

Mahillon, J., Seurinck, J., Van Rompuy, V., Delcour, J. and Zabeau, M. (1985) Nucleotide sequence and structural organization of an insertion sequence element (IS231) from *Bacillus thuringiensis* strain Berliner 1715. *EMBO Journal* 4, 3895–3899.

Martin, P.A.W. and Travers, R.S. (1989) Worldwide abundance and distribution of *Bacillus thuringiensis* isolates. *Applied and Environmental Microbiology* 55, 2437–2442.

Odell, J.T., Nagy, F. and Chua, N.H. (1985) Identification of DNA sequences required for activity of the cauliflower mosaic virus 35S promoter. *Nature* 313, 810–812.

Ohba, M., Yu, Y.M. and Aizawa, K. (1987) Non-toxic isolates of *Bacillus thuringiensis* producing parasporal inclusions with unusual protein components. *Letters in Applied Microbiology* 5, 29–32.

Pao-intara, M., Angsuthanasombat, C. and Panyim, S. (1988) The mosquito larvicidal activity of 130 kDa delta-endotoxin of *Bacillus thuringiensis* var. *israelensis* resides in the 72 kDa amino-terminal fragment. *Biochemical and Biophysical Research Communications* 153, 294–300.

Peferoen, M., Jansens, S., Reynaerts, A. and Leemans, J. (1990) Potato plants with engineered resistance against insect attack. In *Molecular and Cellular Biology of the Potato*, ed. M.E. Vayda and W.C. Park. CAB International, Wallingford, pp. 193–204.

Potrykus, I. (1990) Gene transfer to cereals: an assessment. *Bio/Technology* 8, 535–542.

Sacchi, V.F., Parenti, P., Hanozet, G.M., Giordana, B., Lüthy, P. and Wolfersberger, M.G. (1986) *Bacillus thuringiensis* toxin inhibits K⁺-gradient-dependent amino acid transport across the brush-border membrane of *Pieris brassicae* midgut cells. *FEBS Letters* 204, 213–218.

Schnepf, H.E. and Whiteley, H.R. (1981) Cloning and expression of the *Bacillus thuringiensis* crystal protein gene in *Escherichia coli*. *Proceedings of the National Academy of Sciences (USA)* **78**, 2893–2897.

Schnepf, H.E. and Whiteley, H.R. (1985) Delineation of a toxin-encoding segment of a *Bacillus thuringiensis* crystal protein gene. *Journal of Biological Chemistry* **260**, 6273–6280.

Schnepf, H.E., Wong, H.C. and Whiteley, H.R. (1985) The amino acid sequence of a crystal protein from *Bacillus thuringiensis* deduced from the DNA base sequence. *Journal of Biological Chemistry* **260**, 6264–6272.

Shimamoto, K., Terada, R., Izawa, T. and Fujimoto, H. (1989) Fertile rice plants regenerated from transformed protoplasts. *Nature* **338**, 274–276.

Sick, A., Gaertner, F. and Wong, A. (1990) Nucleotide sequence of a coleopteran-active toxin gene from a new isolate of *Bacillus thuringiensis* subsp. *tolworthi*. *Nucleic Acids Research* **18**, 1305.

Stone, T.B., Sims, S.R. and Marrone, P.G. (1989) Selection of tobacco budworm for resistance to a genetically engineered *Pseudomonas fluorescens* containing the delta-endotoxin of *Bacillus thuringiensis* subsp. *kurstaki*. *Journal of Invertebrate Pathology* **53**, 228–234.

Tabashnik, B.E. (1989) Managing resistance with multiple pesticide tactics: theory, evidence, and recommendations. *Journal of Economic Entomology* **82**, 1263–1269.

Tabashnik, B.E., Cushing, N.L., Finson, N. and Johnson, M.W. (1990) Field development of resistance to *Bacillus thuringiensis* in diamondback moth (Lepidoptera: Plutellidae). *Journal of Economic Entomology* **83**, 1671–1676.

Teeri, T.H., Lehväslaiho, H., Franck, M., Uotila, J., Heino, P., Palva, E.T., Van Montagu, M. and Herrera-Estrella, L. (1989) Gene fusions to *lacZcd* reveal new expression patterns of chimeric genes in transgenic plants. *EMBO Journal* **8**, 343–350.

Thorne, L., Garduno, F., Thompson, T., Decker, D., Zounes, M., Wild, M., Walfield, A.M. and Pollock, T. (1986) Structural similarity between the Lepidoptera- and Diptera-specific insecticidal endotoxin genes of *Bacillus thuringiensis* subsp. *kurstaki* and *israelensis*. *Journal of Bacteriology* **166**, 801–811.

Toriyama, K., Arimoto, Y., Uchimiya, H. and Hinata, K. (1988) Transgenic rice plants after direct gene transfer into protoplasts. *Bio/Technology* **6**, 1072–1074.

Uchimiya, H., Fushimi, T., Hashimoto, H., Harada, H., Syono, K. and Sugawara, Y. (1986) Expression of a foreign gene in callus derived from DNA-treated protoplasts of rice (*Oryza sativa* L.). *Molecular and General Genetics* **204**, 204–207.

Uchimiya, H., Handa, T. and Brar, D.S. (1989) Transgenic plants. *Journal of Biotechnology* **12**, 1–20.

Vaeck, M., Reynaerts, A., Höfte, H., Jansens, S., De Beuckeleer, M., Dean, C., Zabeau, M., Van Montagu, M. and Leemans, J. (1987) Transgenic plants protected from insect attack. *Nature* **327**, 33–37.

Van Rie, J., Jansens, S., Höfte, H., Degheele, D. and Van Mellaert, H. (1990a) Specificity of *Bacillus thuringiensis* δ-endotoxins – importance of specific receptors on the brush border membranes of the mid-gut of target insects. *European Journal of Biochemistry* **186**, 239–247.

Van Rie, J., Jansens, S., Höfte, H., Degheele, D. and Van Mellaert, H. (1990b) Receptors on the brush border membrane of the insect midgut as determinants of the specificity of *Bacillus thuringiensis* delta-endotoxins. *Applied and Environmental Microbiology* **56**, 1378–1385.

Van Rie, J., McGaughey, W.H., Johnson, D.E., Barnett, B.D. and Van Mellaert, H. (1990c) Mechanism of insect resistance to the microbial insecticide *Bacillus thuringiensis. Science* **247**, 72–74.

Velten, J., Velten, L., Hain, R. and Schell, J. (1984) Isolation of a dual plant promoter fragment from the Ti plasmid of *Agrobacterium tumefaciens. EMBO Journal* **12**, 2723–2730.

Waalwijck, C., Dullemans, A.M., Van Workum, M.E.S. and Visser, B. (1985) Molecular cloning and the nucleotide sequence of the Mr 28 000 crystal protein gene of *Bacillus thuringiensis* subsp. *israelensis. Nucleic Acids Research* **13**, 8206–8217.

Wabiko, H., Raymond, K.C. and Bulla, L.A., Jr (1986) *Bacillus thuringiensis* entomocidal protoxin gene sequence and gene product analysis. *DNA* **5**, 305–314.

Ward, E.S. and Ellar, D. (1987) Nucleotide sequence of a *Bacillus thuringiensis* var. *israelensis* gene encoding a 130 kDa delta-endotoxin. *Nucleic Acids Research* **15**, 7195.

Widner, W.R. and Whiteley, H.R. (1989) Two highly related insecticidal crystal proteins of *Bacillus thuringiensis* subsp. *kurstaki* possess different host range specificities. *Journal of Bacteriology* **171**, 965–974.

Widner, W.R. and Whiteley, H.R. (1990) Location of the dipteran specificity region in a lepidopteran-dipteran crystal protein from *Bacillus thuringiensis. Journal of Bacteriology* **172**, 2826–2832.

Wolfersberger, M.G. (1990) The toxicity of two *Bacillus thuringiensis* δ-endotoxins to gypsy moth larvae is inversely related to the affinity of binding sites on midgut brush border membranes for the toxins. *Experientia* **46**, 475–477.

Yamamoto, T. and McLaughlin, R.E. (1981) Isolation of a protein from the parasporal crystal of *Bacillus thuringiensis* var. *kurstaki* toxic to mosquito larva, *Aedes taeniorhynchus. Biochemical and Biophysical Research Communications* **103**, 414–421.

Chapter 7
Potential of Plant-derived Genes in the Genetic Manipulation of Crops for Insect Resistance

Angharad M.R. Gatehouse, Donald Boulter and Vaughan A. Hilder

Department of Biological Sciences, University of Durham, Science Laboratories, South Road, Durham, DH1 3LE, UK

Introduction

Within the plant kingdom there is a very great diversity of secondary metabolic pathways, leading to the accumulation of a very wide range of secondary compounds in different plants. A protective role against various pests, pathogens and competitors for many plant secondary compounds has been established (Table 7.1) and in recent years the utilization of such compounds in crop protection, either by conventional plant breeding or by genetic engineering, has been, and is being, investigated.

The major criteria on which crops have been selected in the past have been based primarily on high and dependable yields, nutritional value (including low mammalian toxicity) and, where necessary, adaptation to certain environmental conditions. As a consequence of these selection pressures and the heavy reliance on monoculture, the co-evolutionary relationships between crop plants and insects have been severely disrupted over the centuries. The result has been that very few cultivated species have retained the degree of resistance exhibited by their wild relatives (Feeny, 1976). A classic example of this is provided by cotton. Because the complex phenolic gossypol is toxic to mammals, so interfering with the utilization of a major part of this crop as an animal feed, lines devoid of this compound have been selected. This selection pressure has resulted in new 'improved' lines which, though more useful for seedcake production, are extremely susceptible to attack by the cotton budworm (*Heliothis virescens*), towards which gossypol is toxic (Berardi and Goldblatt, 1980). A very similar situation has occurred in the production of low-glucosinolate varieties of oilseed rape.

Table 7.1. Plant secondary metabolites exhibiting insecticidal activity.

Compound	Plant source	Reference
Non-protein antimetabolites		
Alkaloids		
2,5-dihydroxymethyl 3,4-dihydroxypyrrolidine (DMDP)	*Lonchocarpus* spp.	Evans *et al.*, 1985
Castanospermine	*Castanospermum australe*	Nash *et al.*, 1986
Non-protein amino acids		
p-aminophenylalanine	*Vigna* spp.	Birch *et al.*, 1986
Terpenoids		
pyrethroids	Compositae	Mann, 1987
juvabione	*Abies balsama*	Mann, 1987
gossypol	*Gossypium hirsute*	Berardi and Goldblatt, 1980
pyrethrin I	*Chrysanthemum cinerariifolium*	Casida, 1973
sesquiterpenoids	*Cyperus iria*	Toong *et al.*, 1988
phytoecdysteroids	Chenopodiaceae, e.g. *Chenopodium alba*	Báthori *et al.*, 1987
	Ajuga remota	Camps, 1991
Rotenoids (isoflavanoids)	*Lonchocarpus salvadorensis*	Birch *et al.*, 1985

Tannins	*Vicia faba*	Boughdad et al., 1986
Polysaccharides		
pectosans	*Phaseolus vulgaris*	Ishii, 1952
heteropolysaccharides	*Phaseolus vulgaris*	Applebaum and Guez, 1972; Gatehouse et al., 1987
Glucosinolates	*Cruciferae*	Erickson and Feeny, 1974
Cyanogenic glycosides	*Lotus corniculatus*	Mann, 1987
Protein antimetabolites		
Lectins	*Phaseolus vulgaris*	Janzen et al., 1976; Gatehouse et al., 1984; Osborn et al., 1986, 1988, 1989
Arcelin	*Phaseolus vulgaris*	Minney et al., 1990
Protease inhibitors	*Vigna unguiculata*	Gatehouse et al., 1979; Hilder et al., 1987; Johnson et al., 1989
α-Amylase inhibitors	*Solanaceae* *Phaseolus vulgaris*	Gatehouse et al., 1987; Ishimoto and Kitamura, 1988

Mankind has sought to redress this balance, to some extent, by a heavy reliance upon the use of chemical insecticides. Despite the expenditure of approximately US $6 billion annually, it is estimated that 37% of all crop production is lost world-wide to pests and diseases, with 13% lost to insects. A major problem encountered with an almost exclusive reliance upon this means of control is the rapid build-up of resistance by insect pests to such compounds. For example, *H. virescens* rapidly developed resistance to the organochloride insecticides used to control it on cotton. It is not uncommon for examples of resistance in a major pest to be noted within the first year of field use (Metcalf, 1986). Furthermore, in the past, the indiscriminate application of pesticides has, in some cases, exacerbated the problem of insect herbivory – where elimination of a wide range of predatory species along with the primary pests has resulted in secondary pests becoming primary pests with even more devastating effects. An example of this has been illustrated by the brown planthopper (*Nilaparvata lugens*), a major field pest of rice (Heinrichs and Mochida, 1983).

Some of the most serious losses of crop production to insects, in human if not economic terms, occur in the developing nations. Sophisticated chemical control agents are often not available to farmers in the Third World for reasons of cost, and in many cases where they are being utilized the measures necessary for their 'safe' use are lacking, resulting in serious consequences for the health of many agricultural workers. Some more 'user-friendly' method of crop protection is clearly desirable.

Although complete elimination of chemical control agents is neither a realistic nor a feasible possibility in the near future, a significant reduction in their use is both necessary and desirable. The enlightened agronomist is, therefore, looking towards an integrated pest control programme comprising a combination of practices, which might include the judicious use of pesticides, crop rotation, field sanitation and the use of pest-free seeds, but above all exploiting inherently resistant plant varieties (Meiners and Elden, 1978).

Breeding crops for resistance offers many advantages over an almost exclusive reliance upon chemicals, the most obvious ones being.

1. It provides season long protection.
2. Insects are always treated at the most sensitive stage.
3. Protection is independent of the weather.
4. It involves no application costs.
5. It protects plant tissues which are difficult to treat using insecticides. For example, chemical insecticides are inadequate at controlling larvae of *Ceutorhyncus assimilis* (pest of oilseed rape) since they attack the developing ovules within the pods.
6. Only crop-eating insects are exposed.
7. The material is confined to the plant tissues expressing it and therefore does not leach into the environment.

8. The active factor is biodegradable and choice of suitable genes/gene products can ensure it is not toxic to man and animals.

9. Consumer acceptability. In recent years there has been much concern over the presence of pesticide residues in food crops; inherently resistant crops should offer the consumer the alternative of produce containing a well-defined and characterized gene product as opposed to unspecified pesticide residues.

10. Considerable financial savings (Figure 7.1). These pest-resistant crops would thus offer economic advantages to the farmer, which should ultimately benefit the consumer.

Plant genetic manipulation as part of a breeding programme can make a significant contribution in the production of insect-resistant crops and offers two major advantages over using conventional plant breeding alone in that it enables the desired gene(s) to be transferred to the recipient plants without the co-transfer of undesirable characteristics, thereby greatly speeding up the development time of new varieties, and also allows the transfer of genes across incompatibility barriers – genes can be introduced from sources which are wholly unavailable to conventional plant breeding. There are, however, some serious constraints upon this new technology, such as the inability to transform and regenerate some specific crop cultivars, the identification and production of 'useful' genes for transfer, and regulatory barriers, to the adoption of transgenic crops. These constraints are being eroded as it is now possible to successfully transform and regenerate a large number of different crop species of importance in the developed and developing world. Also, the number of different 'insecticidal' genes which have been transferred to foreign plant species and which confer varying levels of resistance is gradually increasing.

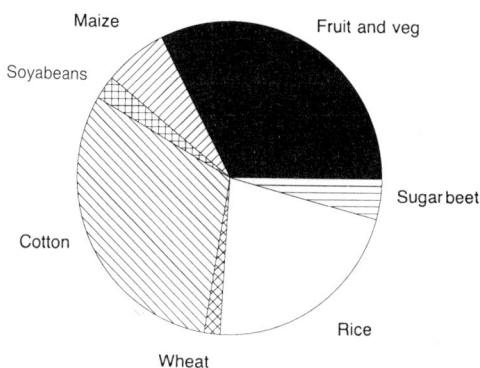

Figure 7.1. Total world insecticide usage, divided according to crops; total: US$ 6.075 billion (1988 figures). Source: County Natwest Woodmac.

This chapter attempts to address the progress to date in the isolation and transformation of insect-resistance genes of plant origin. The strategy employing the use of insecticidal genes of bacterial origin, notably those isolated from *Bacillus thuringiensis*, has been considered in detail in Chapter 6. The reader is also referred to a recent review by Gatehouse, J.A. *et al.* (1991).

1. Potential candidates in engineering insect resistance

1.1 Enzyme inhibitors

Protease inhibitors

The presence of inhibitors of mammalian digestive proteases in plants, particularly those in legume seeds, has been known since 1938 (Read and Haas, 1938) and their incontrovertible demonstration as proteins came in 1945 when Kunitz isolated and crystallized the trypsin inhibitor from soyabeans (Kunitz, 1945). This class of compound has been shown to be very widespread in plants (Richardson, 1977) and those present in the Leguminosae, Graminae and Solananceae have been extensively studied and characterized, presumably because of the large number of species in these families which are important food crops (Richardson, 1991).

The possible role of protease inhibitors in plant protection was investigated as early as 1947 when Mickel and Standish observed that larvae of certain parts were unable to develop normally on soyabean products (Mickel and Standish, 1947). Subsequently the trypsin inhibitors present in soyabean were shown to be toxic to the larvae of the flour beetle, *Tribolium confusum* (Lipke *et al.*, 1954). Following these early studies, there have been many examples of protease inhibitors being active against certain insect species, both in *in vitro* assays against insect gut proteases (Birk *et al.*, 1963; Applebaum, 1964) and in *in vivo* artificial diet bioassays (Steffens *et al.*, 1978; Gatehouse *et al.*, 1979; Broadway and Duffy, 1986; Gatehouse and Hilder, 1988; Burgess *et al.*, 1991). A systematic study of a wide range of different inhibitors has been made in an attempt to find a kinetic parameter which would be useful in predicting the potential of any inhibitor to act as a resistance factor to the larvae of grass grubs (*Costelytra zealandica*), a major insect pest in New Zealand (Christeller and Shaw, 1989); the dissociation constant of the inhibitor:protease complex was suggested as such a parameter. This strategy may well be a valuable tool in choosing the most appropriate enzyme inhibitors to control specific insect pests, though much work will be required to validate this, since food is consumed by organisms, not by enzymes (see Hilder *et al.*, 1991).

Additional evidence for the protective role of protease inhibitors has been provided by Ryan and co-workers (Green and Ryan, 1972) when they demonstrated that damage to the leaves of certain plant species (notably tomato and potato) either by insect feeding or mechanical wounding induces the synthesis of protease inhibitors. This induction occurs not only in the attacked leaf but throughout the plant. Production of these inhibitors was shown to be as the result of a wound hormone, proteinase inhibitor-inducing factor (PIIF), which is released from the damaged leaves and transported to other leaves where it initiates synthesis and accumulation of inhibitors (Shumway *et al.*, 1976; Walker-Simmons and Ryan, 1977; Brown *et al.*, 1985). This evidence strongly implicated protease inhibitors in plant protection (Ryan, 1983). Such a role in protection in the 'field' was first provided by Gatehouse *et al.* (1979) who demonstrated that elevated levels of these inhibitors in one variety of cowpea were partly responsible for the observed resistance of the seeds to the major storage pest of this crop, the bruchid *Callosobruchus maculatus*. This particular trait was later exploited by conventional plant breeding whereby bruchid resistance was transferred to an agronomically improved background (Redden *et al.*, 1983).

The mechanism of antimetabolic action of protease inhibitors is not yet fully elucidated; direct inhibition of digestive enzymes is unlikely to be the only effect and possibly a situation analogous to the effect of some protease inhibitors on mammals, where the major deleterious effect is loss of nutrients through pancreatic hypertrophy and overproduction of digestive enzymes (Liener, 1980), holds in insects. That they affect the nutritional biochemistry of *C. maculatus* has been clearly demonstrated since methionine supplementation has been shown to overcome the antinutritional effects of the cowpea trypsin inhibitor (CpTI) (Gatehouse and Boulter, 1983). Broadway suggests that they are only one part of the complex interaction between plant nutritional value and the insects' digestive physiology (Broadway *et al.*, 1986); this may explain why there are some cowpea varieties with high levels of CpTI which bruchids are able to infest (Xavier-Filho *et al.*, 1989). It is also possible that these compounds may affect other proteases vital for insect development, such as those involved in moulting, and there is preliminary evidence in locust to suggest that they may also be involved in water regulation (A.M.R. Gatehouse, in preparation).

INSECT-RESISTANT TRANSGENIC PLANTS EXPRESSING
PROTEASE INHIBITORS

The first gene of plant origin to be successfully transferred to another plant species resulting in enhanced insect resistance was that isolated from cowpea encoding a trypsin/trypsin inhibitor (Hilder *et al.*, 1987).

This protein was considered to be a particularly suitable candidate for transfer to other plant species via genetic engineering for a number of reasons; it had been shown to be an effective antimetabolite against a range of field and storage pests including members of the Lepidoptera, Coleoptera and Orthoptera (Table 7.2); there was, however, no evidence that it had any deleterious effects upon mammals and it is also a small polypeptide of about 80 amino acids. These inhibitors belong to the Bowman–Birk inhibitor family (Gatehouse *et al.*, 1980) and are encoded by a moderately repetitive gene family in the cowpea genome (Hilder *et al.*, 1989).

A full-length cDNA clone encoding a trypsin/trypsin inhibitor from cowpea was produced and the coding sequence was placed under the control of a CaMV 35S promoter in the final construct produced for transfer to plants (Figure 7.2). The construct employed the *Agrobacterium tumefaciens* Ti plasmid binary vector pROK2; a terminator from the nopaline synthetase gene was placed 3' to the coding sequence, and the construct also contained a nos-neo gene to allow transformants to be selected on the basis of kanamycin resistance. The vector was mobilized into *Agrobacterium*, and the bacteria were used to infect tobacco leaf discs, by standard protocols; subsequent production of rooted plants after

Table 7.2. Insect pests against which cowpea trypsin inhibitors (CpTI) are effective.

Order	Insect pest	Primary crops attacked
Field pests		
Lepidoptera	*Heliothis virescens**	Tobacco, cotton
	*Heliothis zea**	Maize, cotton, beans, tobacco
	Helicoverpa armigera	Cotton, beans, maize, sorghum
	*Spodoptera littoralis**	Maize, rice, cotton, tobacco
	Chilo partellus	Maize, sorghum, sugarcane, rice
	*Autographa gamma**	Sugarbeet, lettuce, cabbage, beans, potato
	*Manduca sexta**	Tomato, tobacco, potato
Orthoptera	*Locusta migratoria*	Polyphagous but preference for wild and cultivated grasses
Coleoptera	*Diabrotica undecimpunctata*	Maize
	Costelytra zealandica	Grasses, clover
	Anthonomus grandis	Cotton
Storage pests		
Coleoptera	*Callosobruchus maculatus*	Cowpea, soyabean
	Tribolium confusum	Most flours

*Insects to which CpTI transgenic tobacco plants exhibit significantly enhanced levels of resistance.

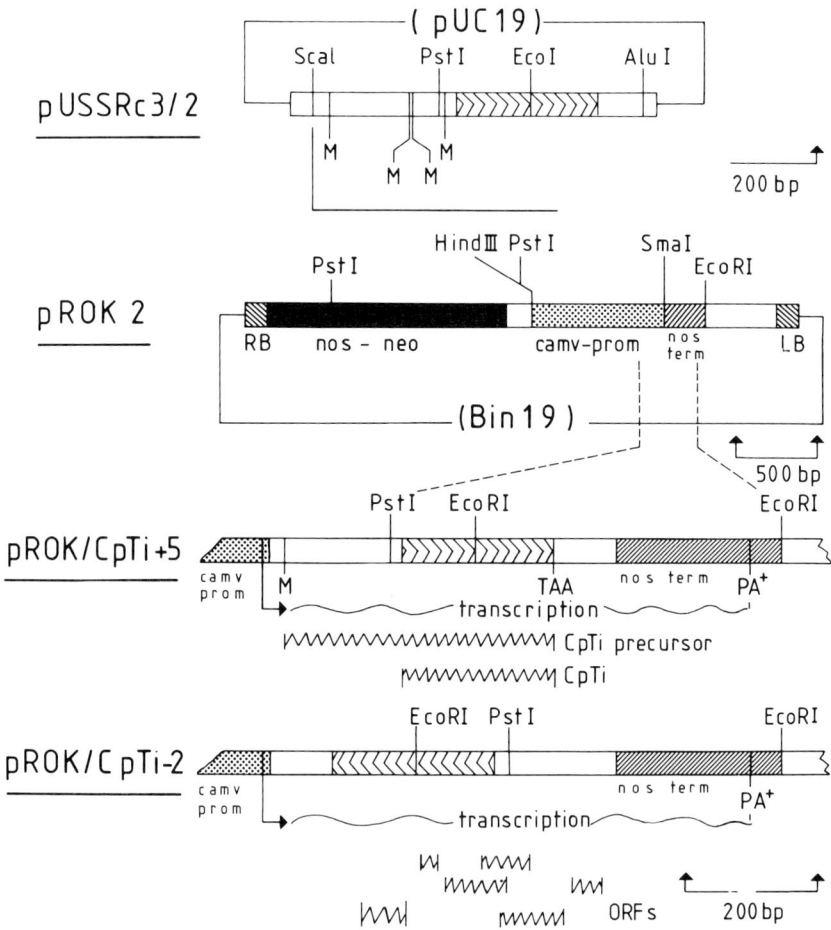

Figure 7.2. Construction of a CpTI expression vector for plant transformation (Hilder *et al.*, 1987). The CpTI cDNA pUSSRc3/2, containing a complete mature CpTI-coding sequence (indicated as > > > ; in frame, initiator codons are marked M), was restricted with Alu I and Sca I and ligated into the Sma I site of the expression vector pROK 2. Clones with the coding sequence in the correct orientation relative to the promoter (pROK/CpTI + 5) and in the incorrect orientation (pROK/CpTI − 2) were generated. Transcripts generated by the clone with the CpTI-coding sequence in the correct orientation will be translated to produce a CpTI precursor polypeptide; transcripts from the clone with the CpTI-coding sequence in the incorrect orientation contain six short open reading frames.

selection of regenerating shoots on kanamycin-containing media also followed normal procedures. By taking cuttings from the original transformants, and rooting them, numbers of clonal plants sufficient for insect bioassay could be produced from each of the original transformants. The transformed plants were shown to be expressing CpTI in the leaves at levels varying from undetectable to nearly 1% of total soluble protein by a dot blot immunoassay; this range of values has subsequently been found to be fairly typical for plant genes driven from the CaMV promoter. In control plants transformed with a construct where the coding sequence of CpTI had been inserted in the incorrect (i.e. 3'–5') orientation relative to the CaMV promoter, no expression of the protein was detected. The expression of CpTI was confirmed in the 'correct' transformants by Western blotting, and by a direct *in vitro* assay for inhibition of bovine trypsin. The former technique showed that tobacco was capable of processing the precursor CpTI polypeptide encoded by the inserted coding sequence to a polypeptide resembling native CpTI on SDS-PAGE; other plant proteins have been shown to be correctly processed in transgenic tobacco plants (Ellis *et al.*, 1988). The latter technique showed that the CpTI synthesized in transgenic tobacco possessed its normal functional integrity. The lack of complication in obtaining relatively high levels of expression of functional CpTI in tobacco illustrates the advantage of expressing plant proteins in transgenic plants; there are no problems with codon usage, mRNA stability, protein processing, etc., which have been observed to occur if attempts are made to express in plants proteins derived from non-plant sources.

The critical test was to ascertain whether the CpTI-producing tobacco plants exhibit enhanced levels of resistance/tolerance to insect infestation compared with the control plants. In the first instance, bioassay of clones of selected transformants was carried out using first instar larvae of the tobacco budworm (*H. virescens*); this insect was chosen as it is a serious pest of tobacco, cotton and maize and thus represents a pest of major economic importance. With these clonal plants, and subsequent generations derived from their self-set seed, the CpTI-expressing plants showed only minor damage compared with the control plants, which in some instances were reduced to stalks (Figure 7.3). Although the larvae begin to feed on the CpTI-expressing plants, causing some limited damage, they either die or fail to develop as they would on control plants. These observations are consistent with a mechanism of CpTI toxicity initially proposed by Gatehouse and Boulter (1983). This protection afforded by CpTI has subsequently been demonstrated for other lepidopteran pests, including *H. zea*, *Spodoptera littoralis* and *Manduca sexta* (Figure 7.4). Statistical analysis of the bioassay in terms of plant damage by leaf area (Figure 7.5a) and insect survival and biomass (Figure 7.5b) confirmed the highly significant protection afforded by CpTI. Recent trials carried out in

Figure 7.3. Bioassay of control and CpTI-expressing transgenic tobacco plants against larvae of *Heliothis virescens* (tobacco budworm). Left, a control showing almost complete destruction; right, a transgenic CpTI-expressor, showing minimal damage.

Figure 7.4. Bioassay of control and CpTI-expressing transgenic tobacco plants against larvae of *Manduca sexta* (tobacco hornworm). Left, a control showing complete destruction; right, a transgenic CpTI-expressor, showing minimal damage.

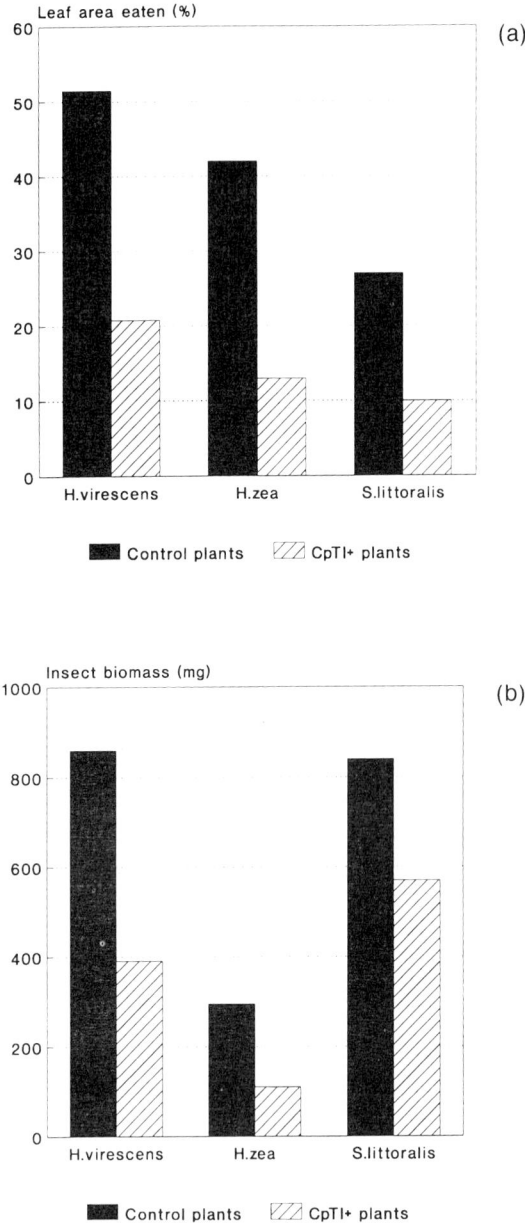

Figure 7.5. Bioassay data for 7-day feeding trials of lepidopteran pests on transgenic tobacco plants expressing CpTI. (a) Leaf area eaten measured by computer-aided image analysis of harvested leaves. (b) Insect biomass. Figure courtesy of Agricultural Genetics Company Ltd.

California showed that expression of CpTI in tobacco afforded significant protection in the field against *H. zea* (Hoffman *et al.*, 1991); results from these trials closely resembled those obtained previously in trials carried out under controlled environmental conditions in growth chambers. Unfortunately, it is not possible to test the efficacy of CpTI against coleopteran pests in transgenic tobacco plants, as most coleopterans of economic interest do not appear to attack tobacco. However, other species of CpTI-expressing plants which are susceptible to coleopteran attack have now become available.

Despite CpTI being an effective antimetabolite against a wide spectrum of insect pests (Table 7.2), recent mammalian feeding trials incorporating the purified protein at levels of 10% of the total protein have failed to demonstrate toxicity (Pusztai *et al.*, 1992). The differences in the organization of the insect and mammalian digestive systems and the vast range of secondary compounds available from plants means that it should not be that difficult to find compounds within plants which are toxic to herbivorous insects but not to mammals (Hilder *et al.*, 1991).

Not only have the genes encoding protease inhibitors isolated from cowpea been shown to confer resistance when expressed in tobacco (Hilder *et al.*, 1987) but the tomato inhibitor II gene, when expressed in the same model plant has also been shown to confer insect resistance (Johnson *et al.*, 1989). The tomato, and potato, inhibitor II gene encodes a trypsin inhibitor (with some chymotrypsin inhibitory activity), and expression of this gene in tobacco on the constitutive promoter CaMV 35S resulted in increased levels of protection against the larvae of the lepidopteran *M. sexta*. These workers showed that the decrease in larval weight was roughly proportional to the level of P I-ll being expressed; at levels of over 100 μg of the foreign protein per g of tissue, larval growth was severely retarded, whereas at lower levels (*c.*50 μg/g tissue) although growth was retarded, it was to a lesser degree. Several of the transgenic plants were shown to contain inhibitor levels of over 200 μg/g tissue; these levels are within the range that is routinely induced by wounding leaves of either tomato or potato plants (Graham *et al.*, 1986).

Interestingly transgenic tobacco plants expressing tomato inhibitor I (a strong inhibitor of chymotrypsin but a weak inhibitor of trypsin) at levels of 130 μg/g supported similar rates of larval growth to those found for control plants. Since P I-II is a trypsin, as opposed to chymotrypsin, inhibitor, it would appear that it is the trypsin inhibitor which is responsible for the observed insecticidal activity. Transformation studies carried out using the CpTI gene involved a gene encoding a trypsin/trypsin inhibitor rather than the trypsin/chymotrypsin inhibitor gene which was also available to us (Hilder *et al.*, 1989).

α-Amylase inhibitors

Protein inhibitors of mammalian α-amylases have been purified and characterized from many plant seeds; they are abundant in cereal grains and are also present in legume and other seeds (Garcia-Olmedo *et al.*, 1987; Richardson, 1991).

Although initially screened against mammalian α-amylases for the presence of inhibitors, plant materials were later screened against crude enzyme preparations from insect guts. This revealed the presence of inhibitors which inhibited both insect and mammalian α-amylases and those specific for the insect enzyme only (Deponte *et al.*, 1976). Screening of the tissues, in particular the seeds, of different plant species against insect gut α-amylases has been carried out in order to identify potential insecticidal genes. However, caution must be given to such an approach since what is effective *in vitro* is not necessarily effective *in vivo*. This is illustrated by the findings of Gatehouse *et al.* (1986) who showed that, although an α-amylase inhibitor from wheat was equally effective in inhibiting gut α-amylase activities from a pest and non-pest species *in vitro*, feeding trials demonstrated it to be relatively innocuous towards the pest species but toxic to the non-pest species (Figure 7.6).

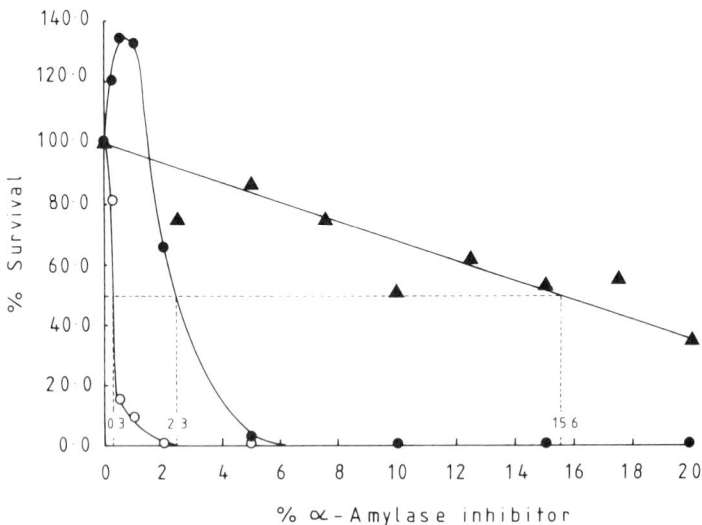

Figure 7.6. Dosage–response curves showing the effects of wheat α-amylase inhibitors on development of *Tribolium confusum* (▲) and *Callosobruchus maculatus* (●). The LC$_{50}$ values were 15.6% and 2.3% respectively. After Gatehouse *et al.* (1986).

Ishimoto and Kitamura (1988) have claimed that an α-amylase inhibitor was responsible for the protection of *Phaseolus vulgaris* seeds against attack by *Callosobruchus chinensis*, a storage pest of certain legumes. They found this inhibitor to be extremely toxic to the larvae, all of which died before the second instar when fed artificial beans containing 0.2%–0.5% of the protein. Birch *et al.* (1989) also demonstrated that a commercially available preparation of *P. vulgaris* α-amylase inhibitor was detrimental to bruchid larval development at concentrations occurring naturally in the seeds. A very specific α-amylase inhibitor has recently been isolated from the seeds of a wild line of *P. vulgaris* (G12953) (Minney *et al.*, 1990) which is resistant to attack by a major storage pest of *P. vulgaris*, *Zabrotes subfasciatus*. Although the purified inhibitor has not as yet been tested in feeding trials against this insect, its presence is correlated with seed resistance to this particular pest species and it has been shown to be a very potent inhibitor of the *Z. subfasciatus* enzyme *in vitro* (Gatehouse *et al.*, 1987). Unlike protease inhibitors, induced synthesis of amylase inhibitors by insect attack has not been observed, and the precise role of these proteins remains in some doubt.

TRANSGENIC PLANTS EXPRESSING α-AMYLASE INHIBITOR

Recently Moreno and Chrispeels (1989) presented strong circumstantial evidence that an α-amylase inhibitor present in the seeds of *P. vulgaris* and active against mammalian and insect, but not plant, α-amylases was encoded by an already identified lectin gene, whose product is referred to as lectin-like protein (LLP). A chimeric gene, consisting of the coding sequence of the lectin gene that encodes LLP and the 5′ and 3′ flanking sequences of the lectin gene that encode phytohaemagglutinin-2, has been made and expressed in transgenic tobacco (Altabella and Chrispeels, 1990). Subsequent analysis of the seeds obtained from these transgenic plants demonstrated the presence of a series of polypeptides (M_r 10000–18000) which cross-reacted with antibodies to the bean α-amylase inhibitor. Since seed extracts from these plants inhibited not only pig pancreatic α-amylase activity but also the α-amylase activity present in the midgut of *Tenebrio molitor* (mealworm), this led the authors to suggest that introduction of this lectin gene (α ai) into other leguminous plants may be a strategy to protect the seeds from the seed-eating larvae of Coleoptera. Although transgenic tobacco plants expressing this gene are available no insect bioassays on these plants appear to have been reported. The authors do, however, express reservations as to the usefulness of this particular inhibitor to protect plants against attack by lepidopteran insects, since the pH optimum for the formation of the protein complex between α-amylase and the inhibitor is 5–6 (Powers and Whitaker, 1977) and it is known that insects in this order have a basic pH, whilst Coleoptera have an acid pH in their midgut (Dow, 1986).

1.2 Lectins

Lectins are carbohydrate-binding proteins found in many plant tissues, and are abundant in the seeds and storage tissues of some plant species (Toms and Western, 1971). The seeds of certain legumes are a particulary rich source of lectins: for example, in *Phaseolus vulgaris* they may be present at levels of up to 3% (Pusztai and Watt, 1974). Many lectins belong to a homologous family of proteins based on an amino acid chain of approximately 220 residues, e.g. soyabean lectin (Lotan *et al.*, 1975), although totally different sequence types have been shown to have similar functional properties.

Toxicity of this type of protein in mammals (Evans *et al.*, 1973; Liener, 1980) and birds (Jayne-Williams and Burgess, 1974) has been well documented and detailed studies on their mechanism of toxicity have been carried out (Pusztai *et al.*, 1979; King *et al.*, 1980a, b; de Oliveira, 1986).

The first report of lectin toxicity towards insects was by Janzen *et al.* in 1976. In this study the purified lectin from *Phaseolus vulgaris* was added to an artificial diet for *C. maculatus* at a range of concentrations from 0.1% to 5.0%. At the highest level there was nò survival and even at 0.1% the lectin was found to cause a significant reduction in the number of larvae developing into adults. Gatehouse *et al.* (1984) subsequenty confirmed the toxicity of the seed lectins of *P. vulgaris* towards developing larvae and on the basis of indirect immunofluorescence investigations using monospecific antisera for globulin lectins showed that the molecules, when ingested, bound to the midgut epithelial cells. From these observations it was suggested that the mechanism of lectin toxicity is analogous to that believed to occur in rat, namely that the ingested lectin causes disruption of the midgut epithelial cells. This leads to a breakdown of nutrient transport, and also facilitates the absorption of potentially harmful substances. In pest species of *P. vulgaris*, such as *Acanthoscelides obtectus*, the intact lectin molecules are unable to bind to these epithelial cells, thereby, presumably, avoiding the harmful effects demonstrated by these lectins (Gatehouse *et al.*, 1989). *P. vulgaris* lectin is composed of a mixture of E type and L type subunits and although the ratio of these subunits does vary, generally the E_2-L_2 form predominates. Interestingly, when either the purified E type or L type subunits on their own were tested in artificial diets, neither had any significant detrimental effects upon development of *C. maculatus* (Table 7.3; Boulter and Gatehouse, 1986). The lectins present in the mature seeds of the winged bean *Psophocarpus tetragonolobus* have also been shown to be involved in seed resistance to non-pest species (Gatehouse A.M.R. *et al.*, 1991).

Not only have certain lectins been identified as being insecticidal compounds by demonstrating a protective role within the seed, as typified by the bean lectins and winged bean lectins (Janzen *et al.*, 1976;

Table 7.3. Effect of *Phaseolus vulgaris* seed lectin (PVu lec) and purified E and L type subunits on development of *Callosobruchus maculatus*.

	Treatment (% wt/wt)		% Survival to adult
PVu lec	0.0	(control)	100.0
	0.5		68.0
	1.0		5.7
	2.0		0.9
PVu lec (L$_4$ type)	0.0	(control)	100.0
	0.5		100.0
	1.0		100.0
	2.0		90.1
PVu lec (E$_4$ type)	0.0	(control)	100.0
	0.5		93.9
	1.0		90.0
	2.0		100.0

Modified from Boulter and Gatehouse (1986).

Gatehouse, A.M.R. *et al.*, 1984, 1991), but several studies have also been carried out involving the screening of purified lectins against insect pests in an attempt to identify insecticidal proteins (Murdock *et al.*, 1990) and hence isolate the genes encoding them for subsequent plant transformation. Shukle and Murdock (1983) demonstrated that the lectin from soyabean was toxic to the developing larvae of the lepidopteran, *Manduca sexta*, when tested in bioassay and recently wheat germ agglutinin has been shown to have an inhibitory effect on development of the European corn borer (*Ostrinia nubilalis*) and the Southern corn root worm (*Diabrotica undecimpunctata*), two important pests of maize (Czapla and Lang, 1990). The disadvantage, however, in using the above-mentioned lectins, namely those from *P. vulgaris*, winged bean, soyabean or wheat germ, is that they have also been shown to exhibit mammalian toxicity (Pusztai *et al.*, 1979; Higuchi *et al.*, 1984). This will therefore limit their potential use in the production of transgenic plants, particularly in crop plants.

The lectin isolated from pea, *Pisum sativum*, on the other hand has been shown to be innocuous to mammals as they are readily broken down in the gut (Begbie and King, 1985). However, they have been shown to be insecticidal when the purified protein was incorporated into artificial seeds (Figure 7.7) (Boulter and Gatehouse, 1986). In addition, those purified from snowdrop and garlic have also been shown to be insecticidal but do not exhibit mammalian toxicity (Pusztai *et al.*, 1992). Of significance is the finding that snowdrop lectin is toxic to the brown planthopper (*Nilaparvata*

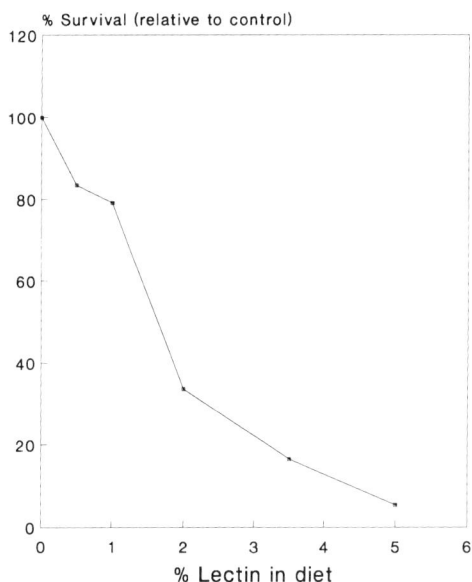

Figure 7.7. Dosage response curve showing the effects of the lectin from *Pisum sativum* on development of *Callosobruchus maculatus*. After Boulter and Gatehouse (1986).

lugens); this is the first evidence of proteins being effective against sap sucking insects.

TRANSGENIC PLANTS EXPRESSING LECTIN

Genes encoding the pea lectin (P-Lec) have been expressed at high levels in transgenic tobacco plants from the CaMV 35S promoter by *Agrobacterium* transformation (Edwards, 1988). P-Lec-expressing plants were then tested in bioassay for enhanced levels of resistance/tolerance to *Heliothis virescens*. The results showed that not only was larval biomass significantly reduced on the transgenic plants, compared with that from control plants (Figure 7.8), but leaf damage, as determined by computer-aided image analysis, was also reduced (Figure 7.8) (Boulter *et al.*, 1990). Transgenic tobacco plants containing both the cowpea trypsin inhibitor gene (CpTI) and the P-Lec gene were obtained by cross-breeding plants derived from the two primary transformed lines. These plants expressing the two insecticidal genes, each at approximately 1% of the total soluble protein, were also screened for enhanced resistance to *H. virescens*. Although the insecticidal effects of the two genes were not synergistic, they were additive (Figure 7.9), with insect biomass on the double expressers being only 11% compared with those from control plants and 50% of those from plants expressing either CpTI or P-Lec alone (Figure 7.9). Leaf damage was also the least on the double-expressing plants (Figure 7.9). Not only is this the first example of a lectin gene being successfully

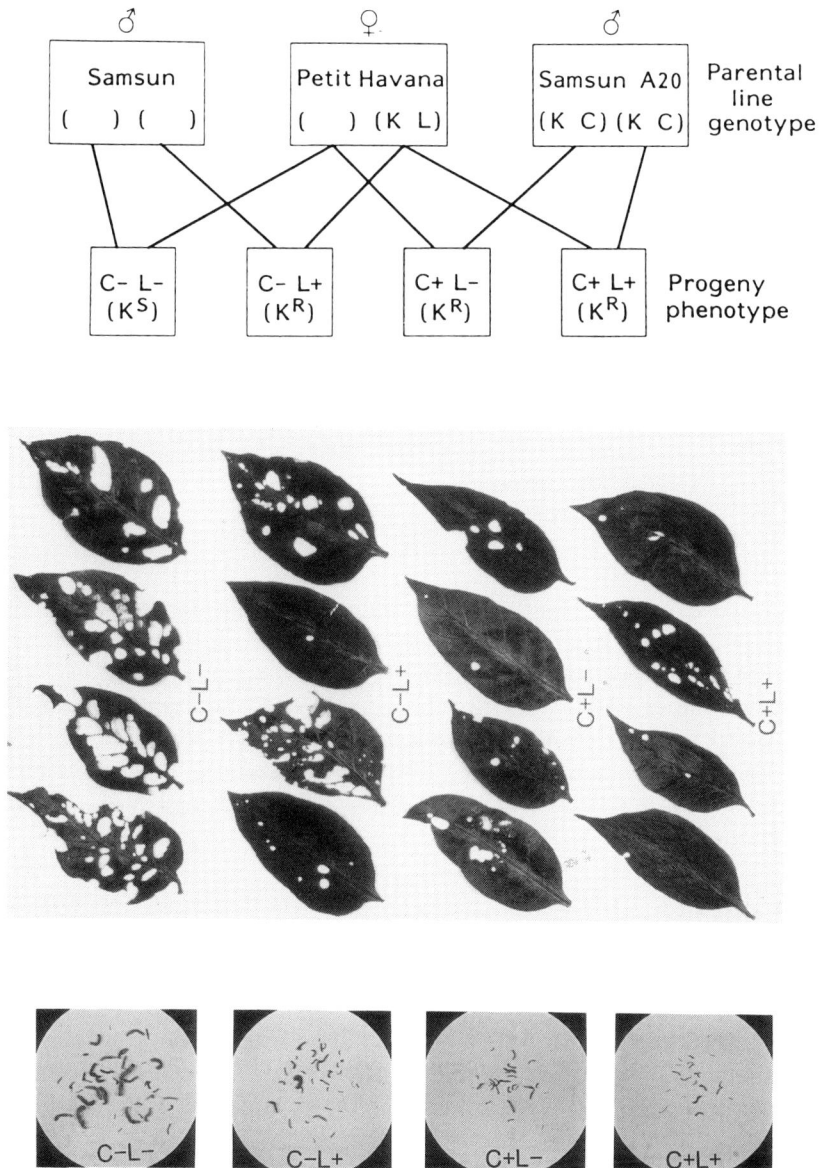

Figure 7.8. Derivation of the four phenotypes in the progeny of crosses between P-Lec- and CpTI-expressing transgenic tobacco plants and between P-Lec and control plants. Leaf damage for the third fully expanded leaf from the apex (the most damaged) and surviving *Heliothis virescens* larvae for each cross are shown. Progeny phenotype: C+, CpTI expression; L+, P-Lec expression; K^R, kanamycin-resistant; K^S kanamycin-susceptible.

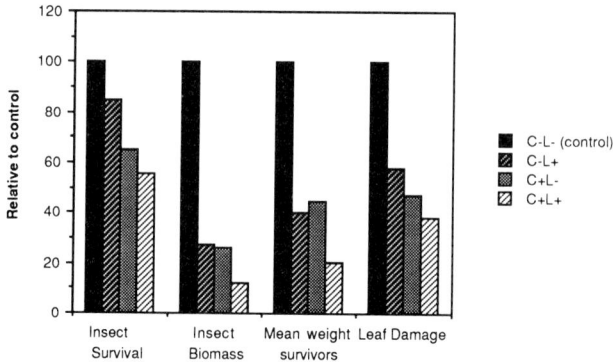

Figure 7.9. Bioassay data showing insect survival and biomass, and leaf damage of CpTI- and P-Lec-expressing transgenic tobacco plants against *Heliothis virescens.* After Boulter *et al.* (1990).

transferred to another plant species resulting in enhanced insect resistance, but it is also the first demonstration of additive protective effects of different plant-derived insect-resistance genes.

The gene encoding the snowdrop lectin has also been successfully transferred to other plants, including tobacco, where preliminary results indicate that it too has conferred a measure of resistance, in terms of both insect biomass and leaf damage (A.M.R. Gatehouse and V.A. Hilder, unpublished data).

1.3 Lectin-like proteins

Arcelin

A novel seed protein (M_r 35000–38000) designated arcelin was found to be present in several resistant wild lines of *Phaseolus vulgaris* (Romero Andreas *et al.*, 1986; Osborn *et al.*, 1988) where it was shown to constitute the major seed storage protein at the expense of phaseolin (Gatehouse *et al.*, 1990). Backcrossing experiments showed a strong correlation between the presence of this protein and resistance to a major storage pest, *Zabrotes subfasciatus* (Osborn *et al.*, 1989) and subsequent feeding trials using the purified protein demonstrated that, when it was added to artificial seeds at levels significantly lower than the physiological concentration present in the wild resistant seed, adult survival was reduced by 52% (Minney *et al.*, 1990). Recently one of the varients of arcelin has been expressed in tobacco but as far as we are aware insect bioassays to investigate the levels of resistance afforded by this protein have not as yet been carried out. Since this protein has been shown to be resistant to digestion, *in vitro*, by insects susceptible to it (Minney *et al.*, 1990) and since, when present, it replaces the major storage protein, it might be anticipated that when expressed in a background of readily digestible proteins, as is required in food crops, it may not prove to confer resistance.

Conclusion

From the work carried out to date, the use of plant-derived insect-resistance genes and transferring them to other plant species by genetic manipulation are a viable means of producing crops with significantly enhanced levels of resistance. The strategy employed by the authors is to produce gene packages whose products are targeted to different bio-chemical and physiological processes within the insect. In this way, it is hoped to provide a multimechanistic form of resistance which can be tailored to the different crops and prevailing insect pests at a given time. Hilder *et al.* (1990) have recently demonstrated that tobacco plants can express plant-derived 'foreign' proteins to moderately high levels with no obvious deleterious effects, and thus this proposed strategy should be possible in other crops.

References

Altabella, T. and Chrispeels, M.J. (1990) Tobacco plants transformed with the bean α ai gene express an inhibitor of insect α-amylase in their seeds. *Plant Physiology* **93**, 805–810.

Applebaum, S.W. (1964) Physiological aspects of host specificity in the bruchidae. I. General considerations of developmental compatibility. *Journal of Insect Physiology* **10**, 783–788.

Appelbaum, S.W. and Guez, M. (1972) Comparative resistance of *Phaseolus vulgaris* beans to *Callosobruchus chinensis* and *Acanthoscelides obtectus* (Col. Bruchidae): the differential digestion of soluble heteropolysaccharide. *Entomologia Experimentalis et Applicata* **25**, 64–74.

Báthori, M., Máthé, J. Jr, Solymosi, P. and Szendrei, K. (1987) Phytoecdysteroids in some species of Caryophyllaceae and Chenopodiaceae. *Acta Botanica Hungarica* **33**, 377–385.

Begbie, R. and King, T.P. (1985) The interaction of dietary lectin with porcine small intestine and the production of lectin-specific antibodies. In *Lectins*, Vol. IV, eds T.C. Bog-Hansen and J. Breborowicz. Walter de Gruyter and Co., Berlin, pp. 15–17.

Berardi, L.C. and Goldblatt, L.A. (1980) Gossypol. In *Toxic Constituents of Plant Foodstuffs*, 2nd edn, ed. I.E. Liener. Academic Press, New York, pp. 183–237.

Birch, A.N.E., Crombie, L. and Crombie, W.M. (1985) Rotenoids of *Lonchocarpus salvadorensis*: their effectiveness in protecting seeds against bruchid predation. *Phytochemistry* **24** (12), 2881–2883.

Birch, A.N.E., Fellows, L.E., Evans, S.V. and Docherty, S.V. (1986) Para-aminophenylalanine in *Vigna*: possible taxonomic and ecological significance as a seed defense against bruchids. *Phytochemistry* **25** (12), 2745–2751.

Birch, A.N.E., Simmonds, M.S.J. and Blaney, W.M. (1989) Chemical interactions between bruchids and legumes. In *Advances in Legume Biology*, eds C.H. Stirton and C.J. Zarucchi. Monographs in Systematic Botany, Missouri Botanic Gardens. Vol. 20, 181–809.

Birk, Y., Gertler, A. and Khalef, S. (1963) Separation of a *Tribolium*-protease inhibitor from soybeans on a calcium phosphate column. *Biochimica et Biophysica Acta* **67**, 326–328.

Boughdad, A., Gillon, Y. and Gagnepain, C. (1986) Influence des tannins condenses due tegument de feves (*Vicia faba*) sur le development larvaire de *Callosobruchus maculatus*. *Entomologia Experimentalis et Applicata* **42**, 125.

Boulter, D. and Gatehouse, A.M.R. (1986) Isolation of genes involved in pest and disease resistance. In *Biomolecular Engineering in the European Community*, ed. E. Magnien. Martinus Nijhoff, Dordrecht, pp. 715–725.

Boulter, D., Edwards, G.A., Gatehouse, A.M.R., Gatehouse, J.A. and Hilder, V.A. (1990) Additive protective effects of incorporating two different higher plant derived insect resistance genes in transgenic tobacco plants. *Crop Protection* **9**, 351–354.

Broadway, R.M. and Duffy, S.S. (1986) Plant proteinase inhibitors: mechanism of action and effect on the growth and digestive physiology of larval *Heliothis zea* and *Spodoptera exigua*. *Journal of Insect Physiology* **32**, 827–833.

Broadway, R.M., Duffey, S.S., Pearce, G. and Ryan, C.A. (1986) Plant proteinase inhibitors: a defense against herbivorous insects? *Entomologia Experimentalis et Applicata* **41**, 33–38.

Brown, W.E., Takio, K., Titani, K. and Ryan, C.A. (1985) Wound-induced trypsin inhibitor in alfalfa leaves: identity as a member of the Bowman–Birk inhibitor family. *Biochemistry* **24**, 2105–2108.

Burgess, E.P.J., Stevens, P.S., Keen, G.K., Laing, W.A. and Christeller, J.T. (1992) Effects of protease inhibitors and dietary protein level on the black field cricket *Teleogryllus commodus*. *Entomologia Experimentalis et Applicata* (in press).

Camps, F. (1991) Plant ecdysteroids and their interactions with insects. In *Annual Proceeedings of the Phytochemical Society of Europe*, ed. J.B. Harborne and F.A. Tomas-Barberan. Clarendon Press, Oxford, pp. 331–376.

Casida, J.E. (ed.) (1973) *Pyrethrum: The Natural Insecticide*. Academic Press, New York and London.

Christeller, J.T. and Shaw, B.D. (1989) The interaction of a range of serine proteinase inhibitors with bovine trypsin and *Costelytra zealandica* trypsin. *Insect Biochemistry* **19**, 233–241.

Czapla, T.H. and Lang, B.A. (1990) Effect of plant lectins on the larval development of European corn borer (Lepidoptera: Pyralidae) and Southern corn rootworm (Coleoptera: Chrysomelidae). *Journal of Economic Entomology* **83**, 2480–2485.

de Oliveira, J.T.A. (1986) Seed lectins: the effects of dietary *Phaseolus vulgaris* lectin on the general metabolism of monogastric animals. Ph.D. thesis, Aberdeen University.

Deponte, R., Parlamenti, R., Petrucci, T., Silano, V. and Tomasi, M. (1976) Albumin α-amylase inhibitor families from wheat flour. *Cereal Chemistry* **53**, 805–820.

Dow, J.A.T. (1986) Insect midgut function. *Advances in Insect Physiology* **19**, 187–328.

Edwards, G.A. (1988) Plant transformation using an *Agrobacterium tumefaciens* Ti-plasmid vector system. Ph.D. thesis, Durham University.

Ellis, J.R., Shirsat, A.H., Hepher, A., Yarwood, J.N., Gatehouse, J.A., Croy, R.R.D. and Boulter, D. (1988) Tissue specific expression of a pea legumin gene in seeds of *Nicotiana plumbaginifolia*. *Plant Molecular Biology* **10**, 203–214.

Erickson, J.M. and Feeny, P. (1974) Sinigrin: a chemical barrier to the black Swallow-tail butterfly *Papilo polyxenes*. *Ecology* **55**, 103–111.

Evans, R.J., Pusztai, A., Watt, W.B. and Bauer, D.H. (1973) Isolation and properties of protein fractions from navy beans (*Phaseolus vulgaris*) which inhibit growth of rats. *Biochimica et Biophysica Acta* **303**, 175–184.

Evans, S.V., Gatehouse, A.M.R. and Fellows, L.E. (1985) Detrimental effects of 2,5-dihydroxymethyl-3,4-dihydroxypyrrolidine in some tropical legume seeds on larvae of the bruchid *Callosobruchus maculatus*. *Entomologia Experimentalis et Applicata* **37**, 257–261.

Feeny, P.P. (1976) Plant apparency and chemical defence. *Recent Advances in Phytochemistry* **10**, 1–40.

Garcia-Olmedo, F., Salcedo, G., Sanchez-Monge, R., Gomez, L., Royo, J. and Carbonero, P. (1987) Plant proteinaceous inhibitors of proteinases and alpha-amylases. In *Oxford Surveys of Plant Molecular and Cell Biology*, Vol. IV, ed. B.J. Miflin. Oxford University Press, Oxford, 275–334.

Gatehouse, A.M.R. and Boulter, D. (1983) Assessment of the anti-metabolic effects of trypsin inhibitors from cowpea (*Vigna unguiculata*) and other legumes on development of the bruchid bettle *Callosobruchus maculatus*. *Journal of the Science of Food and Agriculture* **34**, 345–350.

Gatehouse, A.M.R. and Hilder, V.A. (1988) Introduction of genes conferring insect resistance. In *Proceedings of Brighton Crop Protection Conference*, Vol. 3. Lavenham Press Ltd., Lavenham, Suffolk, UK, pp. 1234–1254.

Gatehouse, A.M.R., Gatehouse, J.A., Dobie, P., Kilminster, A.M. and Boulter, D. (1979) Biochemical basis of insect resistance in *Vigna unguiculata*. *Journal of the Science of Food and Agriculture* **30**, 948–958.

Gatehouse, A.M.R., Gatehouse, J.A. and Boulter, D. (1980) Isolation and characterisation of trypsin inhibitors from cowpea. *Phytochemistry* **19**, 751–756.

Gatehouse, A.M.R., Dewey, F.M., Dove, J., Fenton, K.A. and Pusztai, A. (1984) Effect of seed lectin from *Phaseolus vulgaris* on the development of larvae of *Callosobruchus maculatus*; mechanism of toxicity. *Journal of the Science of Food and Agriculture* **35**, 373–380.

Gatehouse, A.M.R., Fenton, K.A., Jepson, I. and Pavey, D.J. (1986) The effects of α-amylase inhibitors on insect storage pests: inhibition of α-amylase *in vitro* and effects on development *in vivo. Journal of the Science of Food and Agriculture* **37**, 727–734.

Gatehouse, A.M.R., Dobie, P., Hodges, R.J., Meik, J., Pusztai, A. and Boulter, D. (1987) Role of carbohydrates in insect resistance in *Phaseolus vulgaris. Journal of Insect Physiology* **33**, 843–850.

Gatehouse, A.M.R., Shackley, S.J., Fenton, K.A., Bryden, J. and Pusztai, A. (1989) Mechanism of seed lectin tolerance by a major insect storage pest of *Phaseolus vulgaris, Acanthoscelides obtectus. Journal of the Science of Food and Agriculture* **47**, 269–280.

Gatehouse, A.M.R., Minney, H.B., Dobie, P. and Hilder, V.A. (1990) Biochemical resistance to bruchid attack in legumes: investigation and exploitation. In *Bruchids and Legumes: Economics, Ecology and Coevolution*, ed. K. Fujii, A.M.R. Gatehouse, C.D. Johnson, R. Mitchel and T. Yoshida. Kluwer Academic Publishers, Dordrecht, pp. 241–256.

Gatehouse, A.M.R., Howe, D.S., Flemming, J.E., Hilder, V.A. and Gatehouse, J.A. (1991) Biochemical basis of insect resistance in winged bean seeds (*Psophocarpus tetragonolobus*) seeds. *Journal of the Science of Food and Agriculture* **55**, 63–74.

Gatehouse, J.A., Hilder, V.A. and Gatehouse, A.M.R. (1991) Genetic engineering of plants for insect resistance. In *Plant Genetic Engineering*, ed. D. Grierson. Plant Biotechnology Series, Vol. 1, Blackie & Son Ltd., London/Chapman and Hall, New York, pp. 105–135.

Graham, J.S., Hall, G., Pearce, G. and Ryan, C.A. (1986) Regulation of synthesis of proteinase inhibitors I and II mRNAs in leaves of wounded tomato plants. *Planta* **169**, 399–405.

Green, T.R. and Ryan, C.A. (1972) Wound-induced proteinase inhibitor in plant leaves: a possible defense mechanism against insects. *Science* **175**, 776–777.

Heinrichs, E.A. and Mochida, O. (1983) From secondary to major pest status: the case of insecticide-induced rice brown planthopper, *Nilaparvata lugens* resurgence. In *Proceedings of XV Pacific Science Congress*, New Zealand.

Higuchi, M., Tsuchiga, I. and Iwai, K. (1984) Growth inhibition and small intestinal lesions in rats after feeding with isolated winged bean lectin. *Agricultural Biological Chemistry* **48** (3), 695–701.

Hilder, V.A., Gatehouse, A.M.R., Sheerman, S.E., Barker, R.F. and Boulter, D. (1987) A novel mechanism of insect resistance engineered into tobacco. *Nature* **330**, 160–163.

Hilder, V.A., Barker, R.F., Samour, R.A., Gatehouse, A.M.R., Gatehouse, J.A. and Boulter, D. (1989) Protein and cDNA sequences of Bowman–Birk protease inhibitors from the cowpea (*Vigna unguiculata* Walp). *Plant Molecular Biology* **13**, 701–710.

Hilder, V.A., Gatehouse, A.M.R. and Boulter, D. (1990) Genetic engineering of crops for insect resistance using genes of plant origin. In *Genetic Engineering of Crop Plants*, ed. G.W. Lycett and D. Grierson. Butterworths, London, pp. 51–66.

Hilder, V.A., Gatehouse, A.M.R. and Boulter, D. (1991) Transgenic plants for conferring insect tolerance-protease inhibitor approach. In *Transgenic Plants*, Vol. I, ed. S. Kung and R. Wu. Academic Press, New York (in press).

Hoffman, M.P., Zalom, F.G., Smilanick, J.M., Malyj, L.D., Kiser, J., Wilson, L.T.,

Hilder, V.A. and Barnes, W.M. (1991) Field evaluation of transgenic tobacco containing genes encoding *Bacillus thuringiensis* d-endotoxin or Cowpea trypsin inhibitor: efficacy against *Helicoverpa zea* (submitted)

Ishii, S. (1952) Studies on the host preference of cowpea weevil (*Callosobruchus chinensis*). *Bulletin of the National Institute of Agricultural Science (Japan)* 1 (c), 185–256.

Ishimoto, M. and Kitamura, K. (1988) Identification of the growth inhibitor on azuki bean weevil in kidney bean (*Phaseolus vulgaris* L.). *Japanese Journal of Breeding* 38, 367–370.

Janzen, D.H., Juster, H.B. and Liener, I.E. (1976) Insecticidal action of the phytohemagglutinin in black bean on a bruchid beetle. *Science* 192, 795–796.

Jayne-Williams, D.J. and Burgess, C.D. (1974) Further observations on the toxicity of navy beans (*Phaseolus vulgaris*) for Japanese quail (*Coturnix coturnix japonica*). *Journal of Applied Bacteriology* 37, 149–169.

Johnson, R., Narvaez, J., An, G. and Ryan, C.A. (1989) Expression of proteinase inhibitors I and II in transgenic tobacco plants: effects on natural defense against *Manduca sexta* larvae. *Proceedings of the National Academy of Sciences (USA)* 86, 9871–9875.

King, T.P., Pusztai, A. and Clark, E.M.W. (1980a) Immunocytochemical localisation of injested kidney bean (*Phaseolus vulgaris*) lectins in rat gut. *Histochemistry Journal* 12, 201–208.

King, T.P., Pusztai, A. and Clarke, E.M.W. (1980b) Kidney bean (*Phaseolus vulgaris*) lectin-induced lesions in rat small intestine. 3. Ultrastructural studies. *Journal of Comparative Pathology* 92, 357–373.

Kunitz, M. (1945) Crystallization of a trypsin inhibitor from soybean. *Science* 101, 668–669.

Liener, I.E. (1980) *Toxic Constituents of Plant Foodstuffs*, 2nd edn. Academic Press, New York.

Lipke, H., Fraenkel, G.S. and Liener, I.E. (1954) Effect of soybean inhibitors on growth of *Tribolium confusum*. *Journal of Agricultural Food Chemistry* 2, 410–415.

Lotan, R., Cacan, R., Cacan, M., Debray, H., Carter, W.C. and Sharon, N. (1975) On the presence of two types of subunit in soybean agglutinin. *FEBS Letters* 57, 100–103.

Mann, J. (1987) *Secondary Metabolism*. Oxford Chemistry Series (ed. P.W. Atkins, J.S.E. Holker and A.K. Holliday), Clarendon Press, Oxford.

Meiners, J.P. and Elden, T.C. (1978) Resistance to insects and diseases in *Phaseolus*. In *Advances in Legume Science*, ed. R.S. Summerfield and A.H. Bunting. International Legume Conference, Kew, pp. 359–364.

Metcalf, R.L. (1986) The ecology of insecticides and the chemical control of insects. In *Ecological Theory and Integrated Pest Management*, ed. M. Kogan. John Wiley and Sons, New York, pp. 251–297.

Mickel, C.E. and Standish, J. (1947) Susceptibility of processed soy flour and soy grits in storage to attack by *Tribolium castaneum* (Herbst). *University of Minnesota Agricultural Experimental Station Technical Bulletin* 178, 1–20.

Minney, B.H., Gatehouse, A.M.R., Dobie, P., Dendy, J., Cardona, C. and Gatehouse, J.A. (1990) Biochemical bases of seed resistance to *Zabrotes subfasciatus* (bean weevil) in *Phaseolus vulgaris* (common bean): a mechanism for arcelin toxicity. *Journal of Insect Physiology* 36, 757–767.

Moreno, J. and Chrispeels, M.J. (1989) A lectin gene encodes the α-amylase inhibitor of the common bean. *Proceedings of the National Academy of Sciences (USA)* **86**, 7885–7889.

Murdock, L.L., Huesing, J.E., Nielsen, S.S., Pratt, R.C. and Shade, R.E. (1990) Biological effects of plant lectins on the cowpea weevil. *Phytochemistry* **29** (1), 85–89.

Nash, R.J., Fenton, K.A., Gatehouse, A.M.R. and Bell, E.A. (1986) Effects of the plant alkaloid castanospermine as an antimetabolite of storage pests. *Entomologia Experimentalis et Applicata* **42**, 71–77.

Osborn, T.C., Blake, T., Gepts, P. and Bliss, F.A. (1986) Bean arcelin 2. Genetic variation, inheritance and linkage relationships of a novel seed protein of *Phaseolus vulgaris* L., *Theoretical and Applied Genetics* **71**, 847–855.

Osborn, T.C., Burow, M. and Bliss, F.A. (1988) Purification and characterization of arcelin seed protein from common bean. *Plant Physiology* **86**, 399–405.

Osborn, T.C., Alexander, D.C., Sun, S.S.M., Cardona, C. and Bliss, F.A. (1989) Insecticidal activity and lectin homology of arcelin seed protein. *Science* **240**, 207–210.

Powers, J.R. and Whitaker, J.R. (1977) Effect of several experimental parameters on combination of red kidney bean (*Phaseolus vulgaris*) α-amylase inhibitor with porcine pancreatic α-amylase. *Journal of Food Biochemistry* **1**, 239–260.

Pusztai, A. and Watt, W.B. (1974) Isolectins of *Phaseolus vulgaris*. A comprehensive study of fractionation. *Biochimica et Biophysica Acta* **365**, 57–71.

Pusztai, A., Clarke, E.M.W. and King, T.P. (1979) The nutritional toxicity of *Phaseolus vulgaris*. *Nutrition Society* **38**, 115–120.

Pusztai, A., Grant, G., Bardoz, S., Brown, D.J., Stewart, J.C., Ewen, S.W.B., Gatehouse, A.M.R., Hilder, V.A. (1992) Nutritional evaluation of the trypsin inhibitor from cowpea. *British Journal of Nutrition* (in press).

Read, J.W. and Haas, L.W. (1938) Studies on the baking quality of flour as affected by certain enzyme actions. V. Further studies concerning potassium bromate and enzyme activity. *Cereal Chemistry* **15**, 59–68.

Redden, R.J., Dobie, P. and Gatehouse, A.M.R. (1983) The inheritance of seed resistance to *Callosobruchus maculatus* F. in cowpea (*Vigna unguiculata* L. Walp.). I. Analysis of parental, F_1, F_2, F_3 and backcross seed generations. *Australian Journal of Agricultural Research* **34**, 681–695.

Richardson, M.J. (1977) The proteinase inhibitors of plants and microorganisms. *Phytochemistry* **16**, 159–169.

Richardson, M.J. (1991) Seed storage proteins: the enzyme inhibitors. In *Methods in Plant Biochemistry*, Vol. 5, ed. L.J. Rogers. Academic Press, New York, pp. 259–305.

Romero Andreas, J., Yandell, B.S. and Bliss, F.A. (1986) Bean arcelin 1. Inheritance of a novel seed protein of *Phaseolus vulgaris* L. and its effect on seed composition. *Theoretical and Applied Genetics* **72**, 123–128.

Ryan, C.A. (1983) Insect-induced chemical signals regulating natural plant protection responses. In *Variable Plants and Herbivores in Natural and Managed Systems*, ed. R.F. Denno and M.S. McClure. Academic Press, New York, pp. 43–60.

Shukle, R.H. and Murdock, L.L. (1983) Lipoxygenase, trypsin inhibitor, and lectin from soybeans: effects on larval growth of *Manduca sexta* (Lepidoptera: Sphingidae). *Environmental Entomology* **12**, 787–791.

Shumway, L.K., Yang, V.V. and Ryan, C.A. (1976) Evidence for the presence of proteinase inhibitor I in vacuolar protein bodies of plant cells. *Planta* **129**, 161–165.

Steffens, R., Fox, F.R. and Kassel, B. (1978) Effect of trypsin inhibitors on growth and metamorphosis of corn borer larvae *Ostrinia nubilalis* (Hubner). *Journal of Agricultural and Food Chemistry* **26** (1), 170–174.

Toms, G.C. and Western, A. (1971) Phytohaemagglutinins. In *Chemotaxonomy of the Leguminosae*, ed. J.B. Harborne, D. Boulter and B.L. Turner. Academic Press, London, pp. 367–462.

Toong, Y.C., Schooley, D.A. and Baker, F.C. (1988) Isolation of insect juvenile hormone III from a plant. *Nature* **333**, 170–171.

Walker-Simmons, M. and Ryan, C.A. (1977) Immunological identification of proteinase inhibitors I and II in isolated tomato leaf vacuoles. *Plant Physiology* **60**, 61–63.

Xavier-Filho, J., Campos, F.A.P., Ary, M.B., Silva, C.P., Carvalho, M.M.M., Macedo, M.L.R., Lemos, F.J.A. and Grant, G. (1989) Poor correlation between levels of proteinase inhibitors found in seeds of different cultivars of cowpea (*Vigna unguiculata*) and the resistance/susceptibility to predation by *Callosobruchus maculatus*. *Journal of Agricultural and Food Chemistry* **37**, 1139–1143.

Chapter 8
Genetic Engineering of Virus Resistance

Brian Reavy and Michael A. Mayo

Scottish Crop Research Institute, Invergowrie,
Dundee, DD2 5DA, UK

Abbreviations for virus names

ACMV − African cassava mosaic virus; **AlMV** − alfalfa mosaic virus; **BMV** − brome mosaic virus; **BNYVV** − beet necrotic yellow vein virus; **CaMV** − cauliflower mosaic virus; **CMV** − cucumber mosaic virus; **CyRSV** − cymbidium ringspot virus; **GCMV** − grapevine chrome mosaic virus; **PaRSV** − papaya ringspot virus; **PEBV** − pea early browning virus; **PLRV** − potato leafroll virus; **PVS** − potato virus S; **PVX** − potato virus X; **PVY** − potato virus Y; **SHMV** − sunn-hemp mosaic virus; **SMV** − soyabean mosaic virus; **TAV** − tomato aspermy virus; **TBRV** − tomato black ring virus; **TBSV** − tomato bushy stunt virus; **TCV** − turnip crinkle virus; **TEV** − tobacco etch virus; **TMGMV** − tobacco mild green mosaic virus; **TMV** − tobacco mosaic virus; **TobRV** − tobacco ringspot virus; **ToMV** − tomato mosaic virus; **TRV** − tobacco rattle virus; **TSV** − tobacco streak virus; **TVMV** − tobacco vein mottling virus.

Introduction − approaches to achieving virus resistance

Genetic modification of crop plants by the processes of selection and breeding has been used for many years to avoid the effects of virus disease. It has led to some outstanding successes. For example, the Ry gene of *Solanum stoloniferum* is a single major dominant resistance gene giving comprehensive extreme resistance to all strains of potato virus Y (PVY). The Ry gene has been incorporated into several potato (*S. tuberosum*) cultivars and is being used extensively in breeding programmes as a source of resistance to PVY (Ross, 1986). However, there are two major problems in the general application of this approach. First, there may not be a source of a gene conferring resistance to a particular virus and, secondly, even if such a source does exist, there is the need to screen for resistance throughout a lengthy breeding schedule before a homozygous resistant genotype is achieved.

Other naturally occurring mechanisms of protection against virus disease are known. Cross-protection is a phenomenon whereby preinfection of a plant with one strain of a virus producing mild symptoms prevents the development of symptoms caused by other, more virulent, strains of that virus. It has been used successfully to control virus diseases of, for example, tomato (Rast, 1972), citrus fruit trees (Costa and Müller, 1980) and papaya (Yeh and Gonsalves, 1984; Wang *et al.*, 1987). Cross-protection represents in effect one virus nucleotide sequence providing protection against another. The exploitation of genetic transformation technology to achieve expression of virus sequences in plants with the aim of inducing analogous resistance to virus diseases was first suggested by Hamilton (1980). Suggestions of virus sequences the expression of which might mimic cross-protection, have included genes for coat protein (Sequeira, 1984) replicase (Sanford and Johnston, 1985) and also RNA molecules complementary in sequence to virus genes (Palukaitis and Zaitlin, 1984).

Other naturally occurring mechanisms which are known to control virus infections involve the effects of virus-dependent nucleic acid molecules. Satellites exemplify such a molecule. They are small RNA molecules which are found only in association with particular helper viruses and, although they have no appreciable sequence homology with their helper virus, they depend on it for their replication (Murant and Mayo, 1982). The presence of satellite RNAs in some virus cultures modifies the severity of symptoms induced by the infection either by increasing (e.g. Waterworth *et al.*, 1979) or diminishing (Takanami, 1981) the severity of disease symptoms. Less common are defective forms of virus nucleic acid or particles which contain deleted forms of virus genomes and which disrupt the replication of viruses with which they are associated (Holland, 1985). As with cross-protection, attempts have been made to mimic the effects of satellites and DI particles by causing plants to synthesize these types of nucleic acid molecule in order to interfere with virus infections.

Other molecules which have been proposed as potential agents to control virus diseases of plants include ribozymes, which cleave RNA molecules after sequence-specific binding (Haseloff and Gerlach, 1988), and RNA sequences which can direct a ribonuclease to specific RNA molecules (Forster and Altman, 1990).

In this chapter we shall describe in more detail how these approaches have been applied and discuss the future prospects of developing crop plants resistant to virus diseases.

1. Steps in the production of transgenic plants

The steps involved in the production of transgenic plants for virus resistance are the same for all types of antiviral gene. Most usually *Agrobacterium*-mediated transfer is used to deliver the target gene into the
recipient plant. Two main types of vector have been used for constructing
hybrid genes for insertion into plant cells by *Agrobacterium* species (see
Figure 8.1). Co-integrating-type intermediate vectors contain a pBR322
plasmid origin of replication which allows constructs to be produced and
replicated in *Escherichia coli* cells. A transcription cassette, normally a
cauliflower mosaic virus (CaMV) 35S promoter and nopaline synthase

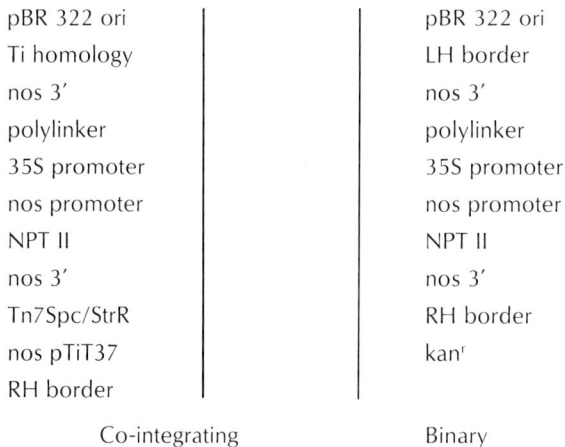

Co-integrating	Binary
pBR 322 ori	pBR 322 ori
Ti homology	LH border
nos 3′	nos 3′
polylinker	polylinker
35S promoter	35S promoter
nos promoter	nos promoter
NPT II	NPT II
nos 3′	nos 3′
Tn7Spc/StrR	RH border
nos pTiT37	kanʳ
RH border	

Figure 8.1. Comparison of structures of co-integrating
and binary Ti plasmid vectors. The vectors are
represented as linear structures. The features indicated
are: pBR322 ori – pBR 322 origin of replication; Ti
homology – region of homology to a Ti plasmid; nos 3′
– 3′ untranslated region of nopaline synthase gene;
polylinker – sequence containing multiple restriction
enzyme sites; 35S promoter – CaMV 35S promoter; nos
promoter – nopaline synthase gene promoter; NPT II –
neomycin phosphotransferase gene conferring
kanamycin resistance to plants; Tn7 Spc/StrR – Tn7
spectinomycin and streptomycin resistance genes; nos
pTiT37 – Ti plasmid T37 nopaline synthase gene; RH
border – T DNA right-hand border sequences; LH
border – T DNA left-hand border sequences; kanʳ –
bacterial kanamycin resistance gene.

(nos) 3′ sequences separated by a polylinker, is included for cloning the target gene. The co-integrating-type intermediate vector also contains a region of homology to Ti plasmids which facilitates recombination between the intermediate vector and a Ti plasmid in *Agrobacterium tumefaciens.* The Ti plasmid is normally 'disarmed', i.e. it lacks a fully functional T-DNA because the right T-DNA border region is missing. Upon recombination, the right T-DNA border is donated by the intermediate vector and a functional T-DNA region containing the target gene is produced. *Agrobacterium*-mediated transformation of plant explants thus transfers the T-DNA region containing the transgenes, that is, the desired gene together with a gene conferring kanamycin resistance, into the plant cell.

In contrast, binary vectors contain both right and left T-DNA borders. Between these borders is a transcription cassette similar to that in the co-integrating-type intermediate vector and a gene conferring kanamycin resistance to plant cells. After cloning in the target gene, the binary vector is transferred into *A. tumefaciens*, where it replicates as an autonomous molecule, utilizing *trans*-acting factors contributed by a helper Ti plasmid harboured by the bacterium. Recombinant *A. tumefaciens* containing the binary vector can infect plant cultures and transfer the genes contained within the T-DNA borders into the nuclear DNA of plant cells in a manner similar to the transfer of co-integrating-type vectors. Transgenic plants are regenerated under kanamycin selection using standard techniques.

2. Coat protein-mediated resistance

2.1 Extent of coat protein-mediated resistance

Most attempts to achieve genetically engineered resistance in plants have used virus coat protein genes in attempts to simulate cross-protection. Typically, a virus coat protein gene is linked to a strong promoter (usually the CaMV 35S promoter) and is stably integrated into the genome of transgenic plants by transfer using agrobacteria as described above. Powell-Abel *et al.* (1986) provided the first demonstration that transgenic plants expressing a virus coat protein gene exhibit resistance to infection with that virus. Transgenic tobacco plants were produced which contained a copy of the tobacco mosaic virus (TMV) strain U_1 coat protein gene. When these plants were inoculated with TMV U_1, symptoms either failed to develop or appeared later than they did in TMV-inoculated control plants. Subsequent characterization of the response of these plants to TMV infection showed that they accumulated less virus than did control plants in both inoculated and systemically infected leaves (Nelson *et al.*, 1987). Since then, virus coat protein genes from a variety of viruses have been introduced into several different types of plants (see Table 8.1) and similar responses to infection with the virus from which the coat protein was derived have been observed (Loesch-Fries *et al.*, 1987; Tumer *et al.*, 1987; van Dun *et al.*, 1987, 1988b; Cuozzo *et al.*, 1988; Hemenway *et al.*, 1988; Nelson *et al.*, 1988; van Dun and Bol, 1988; Hoekema *et al.*, 1989; van den Elzen *et al.*, 1989; Kawchuk *et al.*, 1990; Hill *et al.*, 1991).

Table 8.1. Examples of coat protein-mediated resistance to virus infection.

Source of coat protein gene	Plant transformed	Virus resistance is exhibited to	References
TMV	Tobacco	TMV	Powell-Abel *et al.*, 1986; Nelson *et al.*, 1987
TMV	Tobacco	ToMV, TMGMV	Stark *et al.*, 1990
TMV	Tobacco	PVX, CMV, AlMV, SHMV	Anderson *et al.*, 1989
TMV	Tomato	TMV, ToMV	Nelson *et al.*, 1988
AlMV	Tobacco	AlMV	van Dun *et al.*, 1987, 1988b; Loesch-Fries *et al.*, 1987; Tumer *et al.*, 1987; Halk *et al.*, 1989
AlMV	Tobacco	PVX, CMV	Anderson *et al.*, 1989
AlMV	Tomato	AlMV	Tumer *et al.*, 1987
AlMV	Alfalfa	AlMV	Halk *et al.*, 1989; Hill *et al.*, 1991
TRV	Tobacco	TRV	van Dun and Bol, 1988; Angenent *et al.*, 1990
TRV	Tobacco	PEBV	van Dun and Bol, 1988
TSV	Tobacco	TSV	van Dun *et al.*, 1988b
CMV	Tobacco	CMV	Cuozzo *et al.*, 1988
SMV	Tobacco	PVY, TEV	Stark and Beachy, 1989
BNYVV	Sugarbeet	BNYVV	Kallerhof *et al.*, 1990
PVX	Tobacco	PVX	Hemenway *et al.*, 1988
PVX	Potato	PVX	Hoekema *et al.*, 1989; van den Elzen *et al.*, 1989
PVX + PVY	Potato	PVX, PVY	Lawson *et al.*, 1990; Kaniewski *et al.*, 1990
PVS	Potato	PVS	Mackenzie and Tremaine, 1990
PLRV	Potato	PLRV	Kawchuk *et al.*, 1990
GCMV	Tobacco	TBRV	T. Candresse, personal communication
PaRSV	Tobacco	TEV	Ling *et al.*, 1990
TVMV	Tobacco	TVMV, TEV	Murphy *et al.*, 1990

In addition, coat protein expression has given a measure of resistance to viruses related to that from which the coat protein gene was derived. The development of systemic infection symptoms was delayed in transgenic tomato plants expressing TMV coat protein after infection with tomato mosaic virus (ToMV) (Nelson *et al.*, 1988). Expression of TMV coat protein in tobacco plants also inhibited lesion formation in inoculated leaves and prevented or delayed development of systemic symptoms after infection with the tobamoviruses ToMV and tobacco mild green mosaic virus (TMGMV) (Stark *et al.*, 1990). Delay or total prevention of systemic disease symptoms also occurred in tobacco plants expressing soyabean mosaic potyvirus (SMV) coat protein when they were infected with the potyviruses, PVY or tobacco etch virus (TEV) (Stark and Beachy, 1989). Also, symptoms induced by TEV were slower to develop in tobacco plants expressing the coat protein of tobacco vein mottling potyvirus (TVMV) than in control plants, although there was no difference in virus accumulation. However, in similar tests with TVMV both symptom development and virus accumulation were less in transgenic than in control plants (Murphy *et al.*, 1990). Local lesion formation was unaffected by expression of TMV coat protein when inocula contained viruses unrelated to TMV such as potato virus X (PVX), cucumber mosaic virus (CMV), alfalfa mosaic virus (AlMV) or PVY. However, systemic disease symptoms developed 1–3 days later in transgenic than in control plants when inocula were dilute. A similar delay of systemic disease symptoms happened when tobacco plants expressing AlMV coat protein were inoculated with low concentrations of PVX or CMV (Anderson *et al.*, 1989). Expression of a single virus coat protein gene may therefore be useful in providing differing degrees of resistance to other viruses.

The degree of homology in amino acid sequence between the coat protein expressed by the host and that of the challenge virus required to give protection has not been determined. The coat proteins of the tobamoviruses ToMV and TMGMV mentioned above are 87% and 67% identical to that of TMV. Expression of the SMV coat protein gene protects plants against infection by PVY and TEV (see above), each of which have coat proteins that are about 60% identical to SMV coat protein. Expression of coat protein of one strain of tobacco rattle virus (TRV) did not protect plants against infection by another strain despite the two coat proteins being 39% identical (van Dun and Bol, 1988). However, there was a similar level of homology between coat proteins of TMV and sunn-hemp mosaic virus (SHMV), and TMV coat protein gave plants a low level of protection against SHMV (Stark *et al.*, 1990). Resistance to more than one virus can also be achieved by transformation with the coat protein genes of each virus. Potato plants have been produced which are resistant to both PVX and PVY in this way (Kaniewski *et al.*, 1990; Lawson *et al.*, 1990).

2.2 Effects of inoculum pressure and the amount of coat protein expressed

The effectiveness of protection obtained by transforming plants with coat protein genes can be affected by the amount of coat protein produced in the transgenic plant and by the concentration of virus inoculum used to inoculate the plant in a way which suggests that the protection is dose-dependent. Resistance to infection by TMV, AlMV or PVX by transgenic tobacco plants expressing the homologous coat protein was overcome as the concentration of virus inoculum was increased (Powell-Abel *et al.*, 1986; Tumer *et al.*, 1987; Hemenway *et al.*, 1988; Stark and Beachy, 1989). Figure 8.2 shows an example of this effect in transgenic tobacco plants expressing TMV coat protein. At a low concentration of inoculum (0.4 μg/ml) the onset of disease symptoms was delayed in the transgenic plants to 15 days after inoculation, by which time all the control plants had developed symptoms (Figure 8.2a). Furthermore, only 40% of the transgenic plants showed symptoms 19 days after infection. The resistance in the transgenic plants started to break down when more concentrated inocula were used and symptoms appeared in some inoculated transgenic plants only 5 days after inoculation. While fewer transgenic plants than control plants exhibited symptoms at any time after inoculation, after 19 days almost all the transgenic plants showed symptoms of infection (Figure 8.2b).

The situation is more complex than this, however, and there is a suggestion that the host species as well as the virus genotype may play a role in the dose-dependence. While resistance to TMV infection of transgenic tobacco plants expressing TMV U_1 coat protein was overcome by increasing the concentration of virus inoculum as described above, the resistance of transgenic tomato plants expressing the same coat protein gene was unaffected over a 40-fold range of TMV inoculum concentration, and their resistance to ToMV was also not affected over a 10-fold range of virus concentration (Nelson *et al.*, 1988). Some effect, however, was seen with increasing concentrations of the highly virulent PV230 strain of TMV. Figure 8.3 shows how the complete protection observed with 2 μg/ml of PV230 (Figure 8.3a) was lost when the inoculum concentration was increased to 20 μg/ml (Figure 8.3b). However, the plants still showed some resistance because the onset of symptoms was delayed in the transgenic plants, only 50% of which showed symptoms 29 days after inoculation. As another example, resistance of transgenic tobacco plants expressing AlMV coat protein to infection by AlMV inoculum was overcome at inoculum concentrations of 20 μg/ml (evidenced by an increase in the number of plants showing symptoms in inoculated leaves), whereas transgenic tomato plants expressing the same coat protein were completely resistant when infected with AlMV at a concentration of

Brian Reavy and Michael A. Mayo

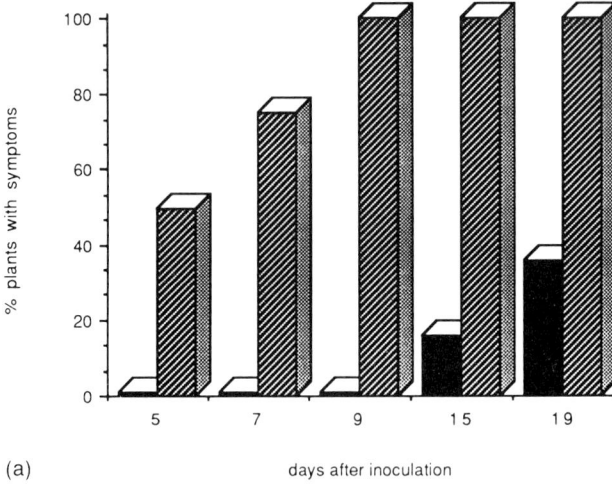

(a)

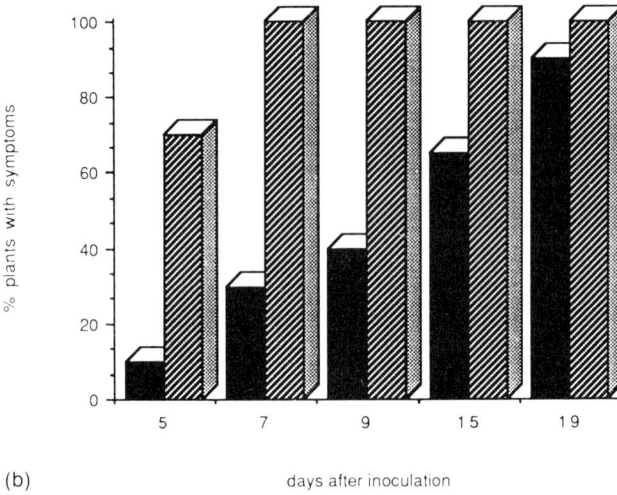

(b)

Figure 8.2. Response of tobacco plants expressing (filled boxes) or not expressing (hatched boxes) TMV U_1 coat protein to infection with TMV U_1 inoculum at 0.4 µg/ml (a) or 2 µg/ml (b). Based on data from Powell-Abel *et al.* (1986).

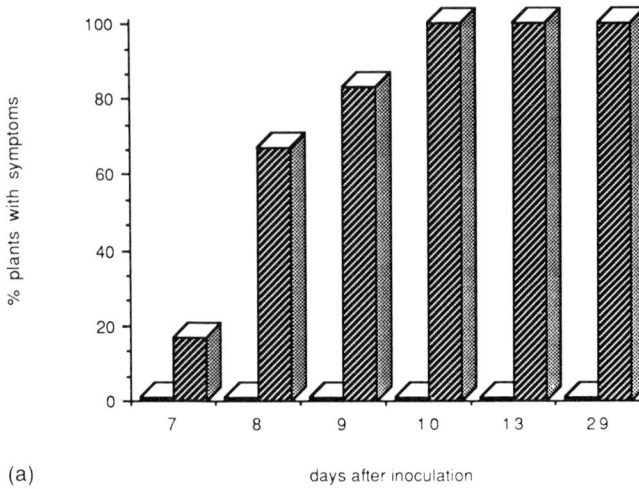

(a)

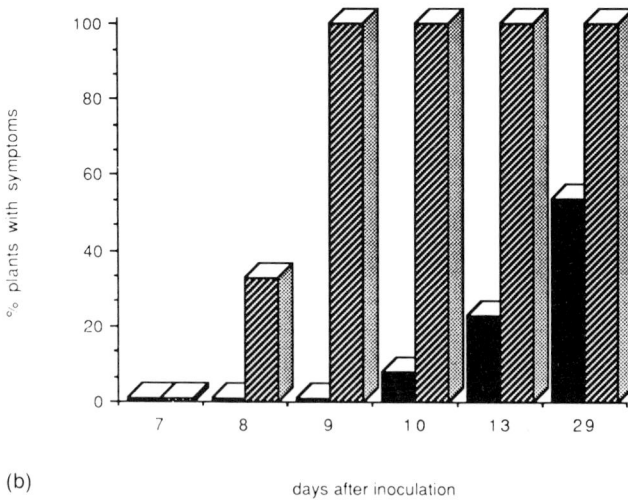

(b)

Figure 8.3. Response of tomato plants expressing (filled boxes) or not expressing (hatched boxes) TMV U_1 coat protein to infection with TMV PV 230 at 2 μg/ml (a) or 20 μg/ml (b). Based on data from Nelson *et al.* (1988).

50 µg/ml (Tumer *et al.*, 1987). It is not clear how these inoculum concentrations delivered by manual inoculation would compare with the inoculum pressure experienced by plants in the field, but potato plants expressing potato leafroll virus (PLRV) coat protein accumulated less virus than control plants when inoculated with 5 or 25 viruliferous aphids/plant (Kawchuck *et al.*, 1990). This may be another example of protection independent of inoculum concentration but it is claimed to represent protection at a higher level of aphid infestation than would be expected in the field. Moreover, this protection occurred in plants containing very little (0.01% of total cell protein) coat protein.

Another possible manifestation of the dose-dependence of coat protein-mediated protection is the effect that the amount of coat protein expression has on resistance. Examples of this are transgenic tobacco, potato and alfalfa plants that expressed low levels of various virus coat proteins and were susceptible to infection by lower inoculum concentrations than plants that expressed more coat protein (Loesch-Fries *et al.*, 1987; Hemenway *et al.*, 1988; Halk *et al.*, 1989; Hoekema *et al.*, 1989; Hill *et al.*, 1991). Similarly, tobacco plants homozygous for an SMV coat protein gene produced higher levels of coat protein and were more resistant than plants heterozygous for the SMV gene (Stark and Beachy, 1989). However, resistance to PVX and PVY infection by transgenic potato plants expressing the coat proteins of both viruses was independent of the level of coat protein expression (Kaniewski *et al.*, 1990; Lawson *et al.*, 1990). Resistance to infection has also been observed against TRV in tobacco (Angenent *et al.*, 1990) and PLRV in potato (H. Barker, A. Kumar, B. Reavy and M. Mayo, unpublished observations) in plants in which no coat protein could be detected, indicating that substantial levels of coat protein expression may not always be required for significant protection. As with the effect of inoculum concentration, the species of plant may be important; the amount of TMV coat protein produced in transgenic tobacco and transgenic tomato plants was decreased when plants were grown at high temperature but, although this treatment led to a loss of resistance to TMV infection in the tobacco plants, resistance of the tomato plants was unaffected (Nejidat and Beachy, 1989).

2.3 Patterns of expression of coat protein in transgenic plants

Nearly all of the transgenic plants that express coat proteins have been produced by transformation using vectors containing the CaMV 35S promoter to direct transcription of the coat protein gene as part of either co-integrating plasmids or binary vectors (see Figure 8.1 for examples). The CaMV 35S promoter is regarded as a strong promoter which is active in a wide range of tissues (Odell *et al.*, 1985; Sanders *et al.*, 1987) and, as would be expected, its use resulted in similar levels of TMV coat protein expression throughout transgenic tobacco plants (Powell-Abel *et al.*, 1986). Differing expression patterns have, however, been observed despite the similarity of the vectors used. Nejidat and Beachy (1989) reported that TMV coat protein levels were highest in older leaves of transgenic tobacco plants. Conversely, AlMV coat protein has been found in larger amounts in young expanding leaves than in older leaves of transgenic tobacco plants (Loesch-Fries *et al.*, 1987) but other workers report either no effect of leaf age upon the amount of AlMV coat protein in transgenic plants (van Dun *et al.*, 1987) or inconsistent variations in expression level (Tumer *et al.*, 1987). Introduction of the PVX coat protein gene into potato plants also gave inconsistent variations in expression level throughout the plant and no coat protein could be detected in these transgenic plants until they were grown in soil (Hoekema *et al.*, 1989). These variable expression patterns could possibly be explained by the effects of different sites of integration of the coat protein genes into the host genomes leading to differing regulation of expression. Another possibility is that the physiological or environmental state of the plant may affect expression of the transgene.

2.4 The action of coat protein in transgenic plants

It has been determined that coat protein-mediated resistance occurs by action of coat protein synthesized in the transgenic plants rather than by some interference effect upon components of the virus replication cycle by the transcript encoding the coat protein. Virus coat protein genes which had been mutated *in vitro* to disrupt synthesis of coat protein were ineffective in inducing resistance to virus infection. Thus, neither, TMV coat protein gene which has a premature stop codon, introduced to produce a truncated protein (Powell *et al.*, 1990), nor an AlMV coat protein gene with a mutated initiation codon, which produced no protein (van Dun *et al.*, 1988b), protected transgenic tobacco plants against virus infection. The ability of virus coat proteins to form aggregates may also be important in inducing resistance to virus infection. Introduction of various TMV coat protein structures into tobacco protoplasts showed that large structures, such as whole UV-inactivated virus particles or rod-like coat protein multimers, gave a longer-lasting and less specific protection against virus infection than did smaller structures such as disc aggregates of coat protein (Register and Beachy, 1989).

2.5 Coat protein-mediated protection operating at a number of different levels

Coat protein-mediated protection, like classical cross-protection between viruses, was overcome by inoculation with purified virus RNA, at least as far as development of symptoms in inoculated leaves is concerned (Loesch-Fries *et al.*, 1987; Nelson *et al.*, 1987; van Dun *et al.*, 1987; Angenent *et al.*, 1990). Exceptions to this occurred with potato plants expressing coat protein of PVX (Hemenway *et al.*, 1988) or *Nicotiana debneyi* plants expressing potato virus S (PVS) coat protein (Mackenzie and Tremaine, 1990). These were resistant to infection with virus RNA. Where resistance is overcome by virus RNA inocula, it seems that an early event in the virus replication cycle, possibly uncoating of infecting virus particles, is responsible for the observed resistance against virus particle inocula. Further evidence supporting the idea that an early event is inhibited comes from studies with protoplasts derived from transgenic tobacco plants that express TMV coat protein. These protoplasts were resistant to infection by TMV and the block to infection occurred before virus RNA synthesis began. Purified virus RNA and partially uncoated TMV particles can both overcome resistance in protoplasts and intact plants, supporting the idea that the initiation of uncoating of virus particles is inhibited (Register and Beachy, 1988). The process of uncoating of virus particles has been examined further in protoplasts derived from transgenic tobacco plants that express TMV U_1 coat protein. A chimeric mRNA encoding the reporter gene, β-glucuronidase (GUS), and containing the origin of assembly sequences from TMV RNA, was packaged by TMV U_1 coat protein to produce pseudovirus particles. The GUS gene is a 'reporter' gene, i.e. one whose activity is easy to assay and which is not found in normal cells. Inoculation of the pseudovirus particles to protoplasts from transgenic plants that express TMV coat protein resulted in only 1–3% of the GUS activity observed in similarly inoculated control protoplasts, confirming that uncoating of virus-like particles is inhibited in the transgenic protoplasts. Furthermore, formation of virus/ribosome complexes thought to be involved in virus uncoating is unhibited in TMV-infected tobacco protoplasts that express TMV coat protein (Wu *et al.*, 1990). However, inhibition of uncoating of virus particles is not the sole mechanism by which coat protein-mediated protection operates. Tobacco plants or protoplasts that expressed TMV U_1 coat protein were not resistant to infection by the distantly related SHMV. Protoplasts that express TMV U_1 coat protein were resistant to infection with TMV U_1 virus particles but were susceptible to infection with hybrid virions consisting of SHMV RNA packaged in TMV U_1 coat protein, even though the results with the GUS reporter gene indicate that 97–99% of virus particles would not be uncoated. This suggests that other effects must inhibit replication of the 1–3% of TMV U_1 virus particles that would be expected to be uncoated in protoplasts expressing TMV U_1 coat protein, but which did not lead to an infection being established (Osbourn *et al.*, 1989).

It has also been shown that, although infection by purified TMV RNA was effective in overcoming resistance in inoculated leaves of transgenic tobacco plants that express TMV coat protein, the development of systemic symptoms was delayed in the transgenic plants compared with control plants (Wisniewski *et al.*, 1990). It is tempting to speculate that the mechanism which delays systemic spread in transgenic tobacco plants infected with TMV RNA is the same as the one which delays the systemic spread of unrelated viruses in the same plants (Anderson *et al.*, 1989). Wisniewski *et al.*, (1990) could find no inhibition of systemic spread upon infection of transgenic tobacco plants expressing TMV coat protein with PVX but the concentration of inoculum used was higher than that used by Anderson *et al.*, (1989). In summary, the two identified mechanisms of coat protein-mediated protection are: (i) prevention of uncoating of virus particles; and (ii) prevention or delay of systemic spread, this possibly being effective against unrelated viruses when the inoculum concentration is low.

3. Attempts to obtain resistance with other virus genes

Coat protein-mediated protection may not mimic all the effects which occur in conventional cross-protection. Experiments with coat protein mutants of TMV that provided cross-protection, even though they produced defective insoluble coat protein or no detectable coat protein at all, suggest that protection mechanisms other than those mediated directly by the coat protein were operating (Zaitlin, 1976; Gerber and Sarkar, 1989). Introduction of a complete cDNA clone of a mild strain of TMV under the control of a CaMV 35S promoter into tobacco plants has been reported to provide protection against infection with unrelated virulent strains of TMV. Unlike coat protein-mediated protection, the protection supplied by the complete cDNA clone was also effective against infection with purified virus RNA (Yamaya *et al.*, 1988).

Individual genes of TMV have also been expressed in plants and their effects upon virus infection determined. Figure 8.4 shows the different genes encoded by TMV and their positions in the virus genome. Introduction of the 54K gene into tobacco provides protection against high levels of both virus (500 µg/ml) and TMV RNA (300 µg/ml) even though no 54K protein could be detected in the transgenic plants (Golemboski *et al.*, 1990). However, transgenic plants that expressed the 126K gene (Golemboski *et al.*, 1990) or the 30K gene (Deom *et al.*, 1987) were not resistant to TMV infection. Attempts to obtain resistance using virus genes other than coat protein have so far been unsuccessful with AlMV and TRV (van Dun *et al.*, 1988a; Angenent *et al.*, 1990). Expression of the TVMV cylindrical inclusion protein (CI) gene in tobacco provided resistance to

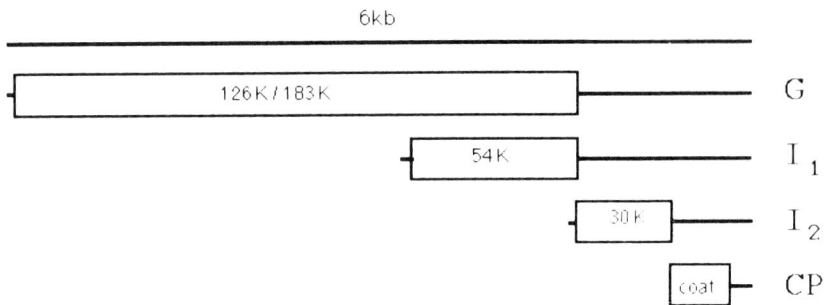

Figure 8.4. Gene organization of TMV. The virus genome is represented as a straight line with the size of 6 kb indicated. The main open reading frames are indicated by open boxes with molecular weights or function of the protein products indicated in the box. Untranslated regions are indicated by a horizontal line.

disease symptom development and virus accumulation following inoculation with TVMV, although the plants themselves displayed a phenotypic abnormality characterized by generalized chlorotic spotting and mottling. Inoculation with TEV of plants expressing the TVMV CI gene induced more severe symptoms than TEV infection of control plants (Murphy *et al.*, 1990).

4. Satellite RNA-mediated resistance

The presence of some kinds of satellite RNA in a virus culture has the effect of decreasing the severity of disease symptoms. The aim of satellite RNA-mediated resistance is to introduce cDNA copies of a satellite RNA into plants so that the satellite RNA is replicated when the target virus infects the transgenic plant, thereby ameliorating the effects of the virus.

The satellite RNA of CMV strain $I_{17}N$ was the first to be introduced into transgenic tobacco plants (Baulcombe *et al.*, 1986). The constructs used consisted of duplicated satellite units consisting of 1.3 or 2.3 satellite RNA molecules in head-to-tail arrangements under the control of a 35S promoter. These constructs were intended to produce RNA molecules which would mimic replicative intermediates thought to occur in plant cells. When tobacco plants that express the $I_{17}N$ satellite were infected with a satellite-free strain (K) of CMV, large amounts of monomeric satellite RNA molecules were produced and symptom development was suppressed, with mosaic and mottling being restricted to the first two to three invaded systemic leaves (Harrison *et al.*, 1987). There was also an 80–95% decrease in the amounts of CMV_K RNA and coat protein produced in the infected transgenic plants and a decrease in the infectivity of leaf extracts, suggesting that CMV replication was inhibited. Similar effects were also seen when the satellite-producing plants were infected with the related virus, tomato aspermy virus (TAV). Infection with TAV led to synthesis of satellite monomers, which became part of the TAV culture, and disease symptoms were modified. However, in contrast to the result with CMV infection, the satellite synthesis during TAV multiplication did not greatly inhibit synthesis of TAV or its RNA (see Table 8.2). There was no effect of any sort on infection of the transgenic plants by unrelated viruses such as brome mosaic virus (BMV), AlMV, tobacco ringspot virus (TobRV), tomato bushy stunt virus (TBSV) or peanut stunt virus. Expression of the CMV satellite has also been shown to be effective in field and glasshouse tests because infection of transgenic pepper plants (Tien *et al.*, 1990b) and transgenic tobacco and tomato plants (Tien *et al.*, 1990a) with CMV resulted in milder disease symptoms and less virus accumulation that in control plants.

In similar work, infection by TobRV was inhibited in transgenic tobacco plants that expressed a naturally occurring TobRV satellite RNA (STobRV) (Gerlach *et al.*, 1987). Trimers of STobRV were introduced into tobacco plants under the control of a CaMV 35S promoter and high levels of monomer satellite RNA were synthesized upon infection with TobRV. In addition, the lesions in inoculated leaves developed 1–2 days later than in control leaves and the lesions lacked a characteristic necrosis. Moreover, there was only a mild systemic reaction on a few leaves some 5–6 weeks later, instead of the severe stunting seen in control plants. Baulcombe *et al.* (1989) suggested that the enhanced suppression of systemic symptoms observed in plants that express the STobRV satellite compared with those that express the CMV satellite may be due to the ability of the STobRV RNA concatemers of self-cleave to produce monomers.

Table 8.2. Virus yields after infection of satellite-transformed plants.

Virus	Yield from satellite-transformed plants (% of control)	
	1/50 Sap dilution	1/250 Sap dilution
CMV	4	5
TAV	125	89

Virus yield was determined by measuring number of lesions on test plants after inoculation with diluted sap from virus-infected satellite-transformed plants. Results are expressed as percentage of lesions observed after inoculation with sap from control infected plants. Data from Harrison *et al.* (1987).

While satellite RNA-mediated resistance is effective in protecting against infection with satellite-free viruses, it is not clear how useful it will be in providing protection against infection with viruses already containing satellites, because satellites can differ in the interactions with different virus strains. Infection of the CMV $I_{17}N$ satellite-transformed tobacco plants with an inoculum containing the Y satellite, along with a strain of CMV with which it is not normally associated, led to amelioration of symptoms on later produced systemically infected leaves and loss of the Y satellite. No effect on symptom development was seen when the Y satellite was mixed with its natural helper strain of CMV and used to infect the transgenic plants, and both satellites were found in the plants (Baulcombe *et al.*, 1989).

5. Defective virus genomic nucleic acid

A similar effect to that of satellite RNA is interference with virus replication by defective interfering RNA molecules. These are deleted forms of virus genome species and they disrupt the replication of viruses with which they are associated. Defective RNA species have been reported in association with TBSV (Hillman *et al.*, 1987), cymbidium ringspot virus (CyRSV) (Burgyan *et al.*, 1989) and turnip crinkle virus (TCV) (Li *et al.*, 1989). Defective forms of BMV RNA 2 have been constructed *in vitro* by deletion mutagenesis of infectious cDNA. These defective RNA 2 molecules depended upon the presence of full-length BMV RNA molecules for replication in protoplasts and specifically interfered with the replication of full-length RNA 2. Infection of protoplasts with equal amounts of RNA 1, RNA 2 and defective RNA 2 resulted in a 37% decrease in the ratio of RNA 2 synthesized relative to RNA 1. Complete interference with the BMV RNA multiplication was observed when the inocula contained five times more defective RNA 2 than RNAs 1 and 2 (Marsh *et al.*, 1990). This suggests that both natural and engineered defective virus genome species may be effective in inducing resistance to virus infection.

African cassava mosaic virus (ACMV) causes a severe mosaic disease of cassava throughout Africa, India and the Indian Ocean, and has a genome consisting of two single-stranded DNA molecules (A and B), each of 2.7 kb (Bock and Harrison, 1985). In addition to these two DNA species, a defective subgenomic DNA derived from the B-DNA by a 50% deletion has been identified. The subgenomic DNA is encapsidated in a manner analogous to defective interfering particles of some other viruses and decreases infectivity when present in ACMV cultures (Stanley and Townsend, 1985). Transgenic *Nicotiana benthamiana* plants containing integrated tandem copies of the subgenomic DNA have been produced and the DNA was excised and replicated when the transgenic plants were infected with ACMV (Stanley *et al.*, 1990). Symptoms in inoculated leaves of these transgenic plants were similar to those in inoculated leaves of control plants but systemic infection was ameliorated in the transgenic plants, which displayed only occasional leaf curling, few chlorotic lesions and little or no stunting instead of the severe leaf curling, extensive chlorosis and marked stunting observed in infected control plants. In addition, less ACMV DNA accumulated in the inoculated transgenic plants than in inoculated control plants. This is the first example of genetically engineered resistance to a plant DNA virus and represents an approach which may be applicable to any virus which has associated defective genomic nucleic acid molecules.

6. Antisense RNA

Antisense RNA in its simplest form can be produced by inverting a cDNA copy of an mRNA with respect to the promoter in an expression vector to give a full-length copy of the complement of an mRNA sequence. Fragments smaller than full-length can also be effective. Antisense RNA molecules are thought to interact with mRNA molecules by base-pairing to form double-stranded RNA. Their presence has been shown to inhibit expression of exogenous and endogenous genes in mammalian cell cultures (Izant and Weintraub, 1985). The first demonstration that expression of antisense RNA could inhibit gene expression in plant cells was with the bacterial chloramphenicol acetyltransferase (CAT) gene as a reporter gene. Vectors were constructed which expressed the CAT gene in either the sense (i.e. mRNA) or antisense orientations. Inhibition of more than 95% of the CAT activity occurred when the sense and antisense RNA-producing vectors were introduced together into carrot protoplasts by electroporation, compared with the sense RNA-producing vector intro-duced on its own (Ecker and Davis, 1986). The amount of inhibition was dependent upon expression of large amounts of the antisense RNA; a 100 : 1 ratio between vectors producing antisense RNA and mRNA-coding vectors was required for inhibition, and the inhibition was greatly reduced if the antisense construct used a relatively weak phenylalanine ammonia lyase gene promoter when the mRNA-coding vector was expressed using a stronger CaMV 35S or nopaline synthase promoter.

Inhibition of expression of endogenous plant genes by expression of antisense RNA leading to observable phenotypic effects is also possible. Probably the most striking of these examples is the use of an antisense RNA complementary to the mRNA encoding chalcone synthase to cause changes in the flower pigmentation pattern of petunia (van der Krol *et al.*,

1988). A variety of flower pigmentation patterns were produced in different transgenic petunia plants that expressed the chalcone synthase antisense RNA and these patterns were stably inherited. Antisense RNA complementary to the mRNA encoding polygalaturonase has been introduced into tomato plants to inhibit the enzyme activity during fruit ripening and this inhibition was also stably inherited (Sheehy *et al.*, 1988; Smith *et al.*, 1988).

Attempts to use antisense RNA molecules to inhibit virus infection of plants were initially disappointing, despite the good results obtained with endogenous genes in plants. Direct comparisons showed that the efficacy of resistance obtained using antisense RNA complementary to virus coat protein genes was less than that obtained with coat protein-mediated protection. Virus accumulation in inoculated leaves was inhibited in tobacco plants that expressed antisense RNA complementary to CMV coat protein sequences, and systemic spread was inhibited at low concentrations of virus inoculum. As with coat protein-mediated protection, the beneficial effects of the CMV antisense RNA were lost as the virus inoculum concentration was increased. The loss of protection with the CMV antisense RNA occurred, however, at low concentrations of virus inoculum and the protective effects of the CMV antisense RNA were lost completely with inoculum concentrations at which coat protein-mediated protection was still effective (Cuozzo *et al.*, 1988). A similar effect of an antisense RNA complementary to the coat protein of PVX has also been reported. Expression of the antisense RNA in tobacco plants inhibited virus accumulation and systemic symptoms upon inoculation with low concentrations of virus but the protection was lost at higher inoculum concentrations, at which coat protein-mediated protection was still effective (Hemenway *et al.*, 1988). These antisense molecules may not be the most effective which could be used. Expression in tobacco plants of an antisense RNA complementary to the TMV coat protein sequence was totally ineffective against infection with TMV. Increasing the length of the expressed antisense RNA so that it included sequences complementary to the 3' end of the virus genome gave a low level of protection at low virus inoculum concentration, suggesting that binding of the antisense RNA to the 3' end of the TMV RNA may prevent replication (Powell *et al.*, 1989). Antisense RNA complementary to the PLRV coat protein gene has also been successful in providing resistance to virus infection (F. van der Wilk, personal communication). Protection against infection by CMV or purified CMV RNA has also been demonstrated in tobacco plants that expressed antisense RNA complementary to the CMV 3a protein gene (Nakayama *et al.*, 1990) but antisense RNA complementary to RNA 2 of TRV gave no resistance to virus infection (Angenent *et al.*, 1990). The efficacy of antisense RNA may therefore be restricted to specific targets or be dependent on expression of large amounts of antisense RNA.

7. Ribozyme-mediated protection

The satellite RNA of tobacco ringspot virus (STobRV) replicates by a rolling circle method which leads to the formation of concatameric molecules. These concatamers are processed to give unit-length STobRV molecules by self-catalysed cleavage at specific sites. The structures responsible for performing the cleavage have been defined and shown to be able to cleave heterologous RNA molecules when flanked by short sequences homologous to these target RNAs (Haseloff and Gerlach, 1988). Specifically, it was shown that the mRNA encoding CAT could be cleaved *in vitro* at specific positions by STobRV catalytic sequences flanked either side by eight bases complementary to the CAT mRNA. These hybrid RNA molecules consisting of satellite RNA endoribonuclease catalytic sequences linked to antisense RNA sequences for specific targeting are called ribozymes.

As well as being able to perform the same function *in vitro*, ribozymes can cleave target RNA molecules *in vivo*. A ribozyme targeted against U7 snRNA molecules was constructed in a tRNA^{met} gene and the chimeric gene introduced into *Xenopus* oocytes by microinjection along with ^{32}P-labelled U7 snRNA. No specific cleavage products were detected, but rather there was an inability to detect any labelled U7 snRNA when there were appreciable amounts of ribozyme sequences in the oocyte cytoplasm, presumably due to degradation of cleavage products (Cotton and Birnstiel, 1989). Large amounts of the ribozyme had to be synthesized in the oocytes to obtain the effect.

In addition to cleavage of RNA, inhibition of gene expression by ribozyme-mediated cleavage of a specific mRNA has also been demonstrated *in vivo*. A 30–60% inhibition of CAT activity in monkey COS1 cells in culture was achieved by simultaneous transient expression of the CAT gene and a ribozyme targeted against it (Cameron and Jennings, 1989). This reduction in enzyme activity was only achieved when the ribozyme sequence was present as a 10^3-fold molar excess over the CAT mRNA. This was done by placing the ribozyme in a replicating vector under the control of the strong simian virus 40 (SV$_{40}$) early promoter and the CAT gene under the control of the relatively weak herpes simplex virus thymidine kinase (TK) promoter. Less ribozyme-mediated suppression of CAT activity occurred when the CAT gene was placed under the control of the stronger human metallothionein gene promoter which gives a 3.5-fold higher CAT expression level than the TK promoter. More pertinent to the possibility of using ribozymes as antiviral agents is the observation that no ribozyme-mediated suppression of CAT activity occurred when the CAT gene was placed in a replicating vector under the control of the strong SV$_{40}$ late promoter giving 8×10^3-fold higher expression than the TK promoter. Although this would apparently question the possible use of ribozymes against virus infections, a ribozyme targeted against the human immuno-deficiency virus 1 (HIV-1) gag gene inhibited HIV-1 infection in cell cultures (Sarver *et al.*, 1990). The ribozyme was placed under the control of the β-actin promoter to produce high levels of transcript and integrated into human cells expressing the HIV receptor, CD4. Specific cleavage of HIV-1 RNA at the gag gene was observed when the ribozyme-expressing cells were infected with HIV-1, with evidence that the infecting virus RNA was being cleaved. The amount of soluble p24 antigen (gag gene product) in infected ribozyme-producing cells was only 1.4–2.3% of that found in cells not producing ribozyme and there was up to a 100-fold decrease in HIV proviral DNA sequences in the ribozyme-producing cells compared with cells not producing ribozyme. There were no apparent cytotoxic effects as a result of producing the ribozyme in the cells, and cell viability and morphology were unaffected as a result of ribozyme production. These results with HIV-1 indicate that ribozyme-mediated resistance can be effective in ameliorating the effects of virus infection and has the potential to be an important method for producing resistance to virus infections. Recently it has been reported that transgenic *Nicotiana tabacum* plants that express ribozymes against TMV show some resistance to TMV infection (Young and Gerlach, 1990).

The identification of a conserved sequence situated 5′ to the initiation site of subgenomic mRNAs among 15 tymoviruses is of particular relevance for preventing virus infections of plants. This conserved sequence is 16 nucleotides long and has been named the tymobox (Ding *et al.*, 1990). The tymobox sequence was used to design a ribozyme that was effective in cleaving various different tymovirus-derived RNAs *in vitro*. Ribozymes could therefore be useful in engendering resistance to several viruses within a group where conserved sequences exist.

Recently, Forster and Altman (1990) have shown that the ribonuclease activity of RNase P can be directed against various target RNA molecules using 'external guide sequences', which are short sequences complementary to a chosen target with a 3′ proximal CCA sequence extension. Binding of external guide sequences to a target RNA creates a partially double-stranded molecule which acts as a substrate for RNase P. This system is a potential alternative to ribozymes but much work has to be done to compare their relative efficacies.

8. Field testing of transgenic resistance

8.1 Cultivar characteristics

It is of major importance in the genetic manipulation of crop plants that cultivar characteristics are unaffected as a result of the procedures used.

Transgenic potato plants that expressed PVX coat protein were tested for 22 morphological and tuber characteristics, including leaf characteristics, anthocyanin coloration, tuber flesh and skin colour, tuber skin smoothness and depth of eyes. Sixty per cent of transgenic Bintje and 87% of Escort transgenic plants were true to type; deviant plants had changes in five or more characteristics (Hoekema *et al.*, 1989). The karyotype of these apparently normal plants was tested and all of 23 transgenic Escort plants and 37 out of 39 transgenic Bintje plants exhibited a normal tetraploid number of chromosomes $(2_n = 48)$. One transgenic Bintje plant was aneuploid $(2_n = 47)$ but phenotypically normal and another was apparently octoploid $(2_n = 96)$ and phenotypically abnormal (van den Elzen *et al.*, 1989). Transgenic tomato plants that expressed TMV coat protein were tested for vegetative traits, such as seed germination percentages, rate of stem elongation, leaf or stem dry matter accumulation and dry matter partioning between leaf and stem, and shown to be normal. Reproductive traits, such as number of flowers per plant, fruit maturation date and fruit yield, were generally unaffected, although fruit yields in one transgenic line were decreased by 65% compared with control plants (Nelson *et al.*, 1988). It is therefore possible to regenerate transgenic plants which display the characteristics of the cultivar from which they are derived, but a close check must be kept for abnormalities which might arise.

8.2 Resistance in the field

Results have been described from field tests of the resistance to virus infection by two types of transgenic plants expressing virus coat proteins. In the first test, 96 transgenic tomato plants that expressed TMV coat protein were planted out in a field and were inoculated with TMV at 39 or 54 days after planting out. Only 3% of inoculated plants that expressed TMV coat protein displayed disease symptoms by the time fruit harvesting began (98 days after planting out), compared with 99% of inoculated non-expressing controls which displayed symptoms. No TMV accumulation was observed in the transgenic plants that displayed no symptoms. Tomato yield from inoculated plants from one coat protein-expressing line was the same as the yields from uninoculated plants that expressed coat protein, or uninoculated non-expressing plants. Tomato yield from inoculated non-expressing plants, on the other hand, was decreased by 27–35% compared with uninoculated control plants (Nelson *et al.*, 1988).

The other field test which has been performed was done on transgenic Russet Burbank potato plants that expressed both PVX and PVY coat proteins. Potato plants were micropropagated *in vitro* and 100 plants of each of four coat protein-expressing lines as well as 100 control Russet Burbank plants were inoculated with PVX and PVY simultaneously. Three days after inoculation the plants were transplanted into a field along with 100 uninoculated transgenic or control plants. Strong disease symptoms, such as severe stunting, leaf discoloration and deformity, were seen in many plants of three of the four inoculated transgenic lines but inoculated plants from one transgenic line (clone 303) showed no obvious disease symptoms. More than 80% of the inoculated plants from the transgenic lines that exhibited symptoms were infected at 16 weeks post-inoculation with either PVX or PVY alone or both together, as was a similar percentage of control Russet Burbank plants. Only 8% of the inoculated plants from clone 303 became infected with PVX or PVY alone and no mixed infections occurred with this clone. Potato tuber yield from inoculated plants of clone 303 was the same as the yield from all four uninoculated transgenic lines or uninoculated untransformed Russet Burbank plants, indicating that the yield from plants of clone 303 was unaffected by virus inoculation. Yields from inoculated plants of the other three transgenic lines were reduced by 45–60% relative to uninoculated plants, compared with a yield reduction of 28% in inoculated untransformed Russet Burbank plants (Kaniewski *et al.*, 1990).

The results of these field trials show that coat protein-mediated protection can be effective in preventing crop losses due to virus infection, but a number of transgenic clones may need to be screened in order to identify the most resistant line.

Field trials have also been performed with plants that expressed CMV satellite RNA. Transgenic tomato plants expressing satellite RNA showed a decrease in disease indices and higher fruit yields upon CMV infection compared with infected control plants. However, no difference in leaf yield was observed between transgenic and control tobacco plants after infection in the field (Tien *et al.*, 1990a).

Both coat protein-mediated and satellite-mediated protection can be very successful in protecting crop yield against the effects of virus infection in the field, and Table 8.3 shows how yields compare between transgenic and normal plants.

9. Future prospects

Table 8.4 summarizes the main features of the different strategies that could be tried to produce virus resistance in transgenic plants. The strategies which are most generally applicable to all plant viruses are coat protein-mediated protection and expression of antisense RNA, ribozymes or targeted RNase P. Coat protein-mediated protection operates at the stage of uncoating of virus particles, a step common to all virus infections, and so may be applicable to DNA viruses as well as the RNA viruses for

Table 8.3. Effect of virus infection on crop yield in transgenic plants.

Field trial	Plant	Yield decrease after infection	Yield increase (% of control)	Reference
TMV coat protein in tomato	Control	26–35% (+)		Nelson *et al.*, 1988
	Transgenic 306–98	0–12% (+)	38–42%	
PVX/PVY coat protein in potato	Control	28%		Kaniewski *et al.*, 1990
	Transgenic 303	0	38%	
CMV satellite in tomato	Transgenic	–	14%	P. Tien, personal communication
	Control inoculated with CMV + satellite control	–	27%	

(+) Depending on time of infection.

which it has been demonstrated. Antisense RNA, ribozymes and the external guide sequences of the RNase P system can all be directed to RNA species which again are common factors in virus infections, whether they are RNA genomes, subgenomic mRNAs or transcripts from DNA viruses. Satellite-mediated protection and defective genome species are not so generally applicable as they are limited in their use to those viruses which are normally found in association with them, and indeed only to the particular viruses which associate with the specific satellite or defective genome species used. For example, the ACMV subgenomic DNA provided resistance to two strains of ACMV but had no effect upon infection with two other geminiviruses, tomato golden mosaic virus or beet curly top virus (Stanley *et al.*, 1990). Defective genome species and satellite-mediated protection may, however, be very effective in those circumstances in which they can be used. The specificity of satellite and defective genome sequences means that they do not provide a broad-spectrum resistance against unrelated viruses, whereas some effect on systemic spread of unrelated viruses was seen with coat protein-mediated protection (Anderson *et al.*, 1989). Antisense RNA, ribozymes and the RNase P system may also be more effective than satellite or defective genome strategies in providing broad-spectrum resistance if conserved sequences useful as targets can be identified among groups of viruses.

A high level of expression of antisense RNA, ribozymes or the external

Table 8.4. Comparison of strategies to obtain virus resistance.

	Coat protein	Satellite	Defective genomes	Antisense RNA	Ribozyme/RNase P
			Type of protection		
Mode of action	Uncoating, virus spread	Interference with replication and symptom development	Interference with replication	Binding to RNA	Cleavage of RNA
General applicability	Yes	No	No	Yes	Yes
Effectiveness against unrelated viruses	Some effect on systemic spread	None	None	Possibly, if conserved sequences can be identified	
Level of expression by host required	Possibly high (?)	Low	Possibly high	Probably high	Possibly high
Efficacy	Can be high	High	Medium	Low	Probably not high
No. of examples	> 20	2	1	4	1

guide sequences of the RNase P system are likely to be required for these methods of protection to be efficacious. This means that screening of large numbers of transgenic plants would probably be required to identify those producing the largest amounts of product of the incorporated transgene. It is not clear what level of expression is required for effective coat protein-mediated protection and it is possible that the effective dose of coat protein may differ in different plant species. Satellite-mediated protection does not require a high level of expression of the satellite in the transgenic plant, as large amounts of monomeric satellite RNA molecules are produced when the plants are infected with virus (Baulcombe *et al.*, 1986; Gerlach *et al.*, 1987). The defective genome species described by Stanley *et al.* (1990) does not require any expression, as the integrated subgenomic DNA is excised and amplified as a result of virus infection.

Satellite-mediated protection has another possible advantage in that the satellite RNA synthesized in the infected plants is incorporated into the virus culture and passed on with the virus when used to infect fresh plants (Harrison *et al.*, 1987), indicating that a virus could possibly be attenuated by passage through transgenic plants that express satellite RNA molecules. The effects of satellite RNAs vary, depending on the host plant they are in, so that satellites which ameliorate symptoms in one plant species may exacerbate them in others. This means that acquisition of a satellite RNA by a virus after passage through a satellite RNA-producing plant may be beneficial in attenuating the effects of the virus in one plant species but may have deleterious effects in others. In order to overcome this problem it may be possible to construct defective satellite molecules that are unable to transfer from the transgenic plant, or satellites that are benign in all species could be developed. Problems also exist concerning the genetic stability of the satellite RNA. It has recently been demonstrated that a single base change may be enough to change the effects of a CMV satellite (WL1-sat) from causing amelioration of symptoms on tomato to causing necrosis (Sleat and Palukaitis, 1990). Recombination between satellite RNA molecules can also occur in plants (Gascone *et al.*, 1990), implying that infection of a satellite RNA-producing transgenic plant with a virus carrying another satellite could lead to the production of novel satellite RNA species. The ACMV subgenomic DNA can also be incorporated into a virus culture after passage through transgenic plants but is not very effective in attenuating the virus culture (Stanley *et al.*, 1990).

Resistance to more than one virus in a single plant has been achieved by expression of PVX and PVY coat protein genes in potato plants (Kaniewski *et al.*, 1990; Lawson *et al.*, 1990). Another way to obtain multivalent resistance is to exploit heterologous proteins. Thus expression of SMV coat protein can provide resistance to several potyviruses (Stark and Beachy, 1989) and expression of TMV coat protein provides resistance to related tobamoviruses (Stark *et al.*, 1990). It is likely that important conserved structural features are involved in these multivalent resistances and it is possible to envisage the development of synthetic genes containing resistance determinants from several virus groups to give a broad range of virus resistances. The observation that mutation of the second amino acid of the AlMV coat protein abolishes protection in transgenic plants (Tumer *et al.*, 1990) is a first step in delineating the important structures involved in coat protein-mediated protection. In addition, combinations of the various resistance strategies, e.g. coat protein and satellite, within a single plant may produce better resistance. For example, tobacco plants have been produced that express both CMV coat protein and CMV satellite RNA. These plants exhibited the inhibition of lesion development on inoculated leaves typical of coat protein-mediated resistance and the inhibition of systemic spread of infection typical of satellite-mediated resistance (Harrison and Murant, 1988).

Other novel antivirus strategies probably still remain to be exploited. Expression of antibodies in transgenic plants has been described (Hiatt *et al.*, 1989) and they may be effective in interfering with virus infections. Endogenous plant genes which confer resistance to virus infections probably still await identification and exploitation. Elucidation of the structures of virus-encoded proteins such as polymerases and proteases could eventually facilitate the design of novel polypeptides which could bind to them and inhibit their functions.

In summary, the first transgenic virus-resistant crop plants utilizing coat protein-mediated resistance or satellite-mediated resistance are becoming available 6 years after the phenomena were first described. Other resistance strategies based upon interference with virus nucleic acids have been developed and are being assessed for their effectiveness and applicability. Further advances will probably lead to the development of other novel resistance strategies. In all probability, improvements in transformation and regeneration technology will allow the extension of these techniques to cereal crops, which are refractory to transformation by agrobacteria.

References

Anderson, E.J., Stark, D.M., Nelson, R.S., Powell, P.A., Tumer, N.E. and Beachy, R.N. (1989) Transgenic plants that express the coat protein genes of TMV and AlMV interfere with disease development of some non-related viruses. *Phytopathology* **79**, 1284–1290.

Angenent, G.C., Van den Ouweland, J.M.W. and Bol, J.F. (1990) Susceptibility to virus infection of transgenic plants expressing structural and non-structural genes of tobacco rattle virus. *Virology* **175**, 191–198.

Baulcombe, D.C., Saunders, G.R., Bevan, M.W., Mayo, M.A. and Harrison, B.D.

(1986) Expression of biologically active viral satellite RNA from the nuclear genome of transformed plants. *Nature (London)* **321**, 446–449.

Baulcombe, D., Devic, M., Jaegle, M. and Harrison, B.D. (1989) Control of viral infection in transgenic plants by expression of satellite RNA of cucumber mosaic virus. In *Molecular Biology of Plant–Pathogen Interactions,* ed. B. Stasawicz, P. Ahlquist and O. Yoder. Alan R. Liss Inc., New York, 257–267.

Bock, K.R. and Harrison, B.D. (1985) *African Cassava Mosaic Virus.* AAB Descriptions of Plant Viruses No. 297 CMI/AAB, Kew, UK.

Burgyan, J., Grieco, F. and Russo, M. (1989) A defective interfering RNA molecule in cymbidium ringspot virus infections. *Journal of General Virology* **70**, 235–239.

Cameron, F.H. and Jennings, P.A. (1989) Specific gene suppression by engineered ribozymes in monkey cells. *Proceedings of the National Academy of Sciences (USA)* **86**, 9139–9143.

Costa, A.S. and Müller, G.W. (1980) Tristeza control by cross protection: a US Brazil cooperation success. *Plant Disease* **64**, 538–541.

Cotten, M. and Birnstiel, M.L. (1989) Ribozyme mediated destruction of RNA *in vivo. EMBO Journal* **8**, 3861–3866.

Cuozzo, M., O'Connell, K.M., Kaniewski, W., Fang, R-X., Chua, H. and Tumer, N.E. (1988) Viral protection in transgenic tobacco plants expressing the cucumber mosaic virus coat protein or its antisense RNA. *Bio/Technology* **6**, 549–557.

Deom, C.M., Oliver, M.J. and Beachy, R.N. (1987) The 30-kilodalton gene product of tobacco mosaic virus potentiates virus movement. *Science* **237**, 389–394.

Ding, S., Howe, J., Keese, P.D., Mackenzie, A., Meek, D., Keese, M.O., Skotnicki, M., Srifah, P., Torronen, M. and Gibbs, A. (1990) The tymobox, a sequence shared by most tymoviruses: its use in molecular studies of tymoviruses. *Nucleic Acids Research* **18**, 1181–1187.

Ecker, J.R. and Davis, R.W. (1986) Inhibition of gene expression in plant cells by expression of antisense RNA. *Proceedings of the National Academy of Sciences (USA)* **83**, 5372–5376.

Forster, A.C. and Altman, S. (1990) External guide sequences for an RNA enzyme. *Science* **249**, 783–786.

Gascone, P.J., Carpenter, C.D., Li, X.H. and Simon, A.E. (1990) Recombination between satellite RNAs of turnip crinkle virus. *EMBO Journal* **9**, 1709–1715.

Gerber, M. and Sarkar, S. (1989) The coat protein of tobacco mosaic virus does not play a significant role for cross-protection. *Journal of Phytopathology* **124**, 323–331.

Gerlach, W.L., Llewellyn, D. and Haseloff, J. (1987) Construction of a plant disease resistance gene for the satellite RNA of tobacco ringspot virus. *Nature* **328**, 802–805.

Golemboski, D.B., Lomonossoff, G.P. and Zaitlin, M. (1990) Plants transformed with a tobacco mosaic virus non-structural gene are resistant to the virus. *Proceedings of the National Academy of Sciences (USA)* **87**, 6311–6315.

Halk, E.L., Merlo, D.J., Liao, L.W., Jarvis, N.P., Nelson, S.E., Krahn, K.J., Hill, K.K., Rashka, K.E. and Loesch-Fries, L.S. (1989) Resistance to alfalfa mosaic virus in transgenic tobacco and alfalfa. In *Molecular Biology of Plant–Pathogen*

Interactions, ed. B.J. Staskawicz, P. Ahlquist and O. Yoder. Alan R. Liss Inc., New York, pp. 283–296.

Hamilton, R.I. (1980) Defenses triggered by previous invaders: viruses. In *Plant Disease: An Advanced Treatise*, Vol. 5, ed. J.G. Horsfall and E.B. Cowling. Academic Press, New York, pp. 279–303.

Harrison, B.D. and Murant, E.A. (1988) Genetic engineering of virus resistance. In *Annual Report of Scottish Crop Research Institute*. pp. 164–166.

Harrison, B.D., Mayo, M.A. and Baulcombe, D.C. (1987) Virus resistance in transgenic plants that express cucumber mosaic virus satellite RNA. *Nature* **328**, 799–802.

Haseloff, J. and Gerlach, W.L. (1988) Simple RNA enzymes with new and highly specific endoribonuclease activities. *Nature* **334**, 585–591.

Hemenway, C., Fang, R-X., Kaniewski, W.K., Chua, N-H. and Tumer, N.E. (1988) Analysis of the mechanism of protection in transgenic plants expressing the potato virus X coat protein or its antisense RNA. *EMBO Journal* **7**, 1273–1280.

Hiatt, A., Cafferkey, R. and Bowdish, K. (1989) Production of antibodies in transgenic plants. *Nature* **342**, 76–78.

Hill, K.K., Jarvis-Eagen, N.J., Halk, E.L., Krahn, K.J., Liao, L.W., Mathewson, R.S., Merlow, D.J., Nelson, S.E., Rashka, K.E. and Loesch-Fries, L.S. (1991) The development of virus-resistant alfalfa, *Medicago sativa* L. *Bio/Technology* **9**, 373–377.

Hillman, B.I., Carrington, J.C. and Morris, T.J. (1987) A defective interfering RNA that contains a mosaic of plant virus genome. *Cell* **51**, 427–433.

Hoekema, A., Huisman, M.J., Molendijk, L., van den Elzen, P.J.M. Cornelissen, B.J.C. (1989) The genetic engineering of two commercial potato cultivars for resistance to potato virus X. *Bio/Technology* **7**, 273–278.

Holland, J.J. (1985) Generation and replication of defective viral genomes. In *Virology*, ed. B.N. Fields. Raven Press, New York, pp. 77–99.

Izant, J.G. and Weintraub, H. (1985) Constitutive and conditional suppression of exogenous and endogenous genes by anti-sense RNA. *Science* **229**, 345–352.

Kallerhoff, J., Perex, P., Bouzoubaa, S., Ben Tahar, S. and Perret, J. (1990) Beet necrotic yellow vein virus coat protein-mediated protection in sugarbeet (*Beta vulgaris* L) protoplasts. *Plant Cell Reports* **9**, 224–228.

Kaniewski, W., Lawson, C., Sammons, B., Haley, L., Hart, J., Delannay, X. and Tumer, N.E. (1990) Field resistance of transgenic Russet Burbank potato to effects of infection by potato virus X and potato virus Y. *Bio/Technology* **8**, 750–754.

Kawchuk, L.M., Martin, R.R. and McPherson, J. (1990) Resistance in transgenic potato expressing the potato leafroll virus coat protein gene. *Molecular Plant-Microbe Interactions* **3**, 301–307.

Lawson, C., Kaniewski, W., Haley, L., Rozman, R., Newell, C., Sanders, P. and Tumer, N.E. (1990) Engineering resistance to multiple virus infection in a commercial potato cultivar: resistance to potato virus X and potato virus Y in transgenic Russet Burbank potato. *Bio/Technology* **8**, 127–134.

Li, X., Heaton, L.A., Morris, T.J. and Simon, A.E. (1989) Turnip crinkle virus defective interfering RNA intensify symptoms and are generated *de novo*. *Proceedings of the National Academy of Sciences (USA)* **86**, 9173–9177.

Ling, K., Namba, S., Gonsalves, C., Slightom, J.L. and Gonsalves, D. (1990) Cross-protection against tobacco etch virus with transgenic tobacco plants expressing the papaya ringspot virus coat protein gene. *Abstracts of VIIIth International Congress of Virology*, Berlin, August 24–26 1990, p. 456.

Loesch-Fries, L.S., Merlo, D., Zinnen, T., Burhop, L., Hill, K., Krahn, K., Jarvis, N., Nelson, S. and Halk, E. (1987) Expression of alfalfa mosaic virus RNA4 in transgenic plants confers virus resistance. *EMBO Journal* 7, 1845–1851.

Mackenzie, D.J. and Tremaine, J.H. (1990) Transgenic *Nicotiana debneyii* expressing viral coat protein are resistant to potato virus S infection. *Journal of General Virology* 71, 2167–2170.

Marsh, L.E., Pogue, G.P. and Hall, T.C. (1990) Artificially derived defective interfering RNAs of brome mosaic virus. *Abstracts of VIIIth International Congress of Virology*, Berlin, August 24–26 1990, p. 123.

Murant, A.F. and Mayo, M.A. (1982) Satellites of plant viruses. *Annual Reviews of Phytopathology* 20, 49–70.

Murphy, J.F., Hunt, A.G., Rhoads, R.E. and Shaw, J.G. (1990) Expression of potyviral genes in transgenic tobacco plants. *Abstracts of VIIIth International Congress of Virology*, Berlin, August 24–26 1990, p. 471.

Nakayama, M., Yoshida, T., Okuno, T. and Furusawa, I. (1990) Protection against cucumber mosaic virus and its RNA infection in transgenic tobacco plants expressing coat protein and antisense RNA of the virus. *Abstracts of VIIIth International Congress of Virology*, Berlin, August 24–26 1990, p. 457.

Nejidat, A. and Beachy, R.N. (1989) Decreased levels of TMV coat protein in transgenic tobacco plants at elevated temperatures reduce resistance to TMV infection. *Virology* 173, 531–538.

Nelson, R.S., Powell-Abel, P. and Beachy, R.N. (1987) Lesions and virus accumulation in inoculated transgenic tobacco plants expressing the coat protein gene of tobacco mosaic virus. *Virology* 158, 126–132.

Nelson, R.S., McCormick, S.M., Delannay, X., Dube, P., Layton, J., Anderson, E.J., Kaniewska, M., Proksch, R.K., Horsch, R.B., Rogers, S.G., Fraley, R.T. and Beachy, R.N. (1988) Virus tolerance, plant growth, and field performance of transgenic tomato plants expressing coat protein from tobacco mosaic virus. *Bio/Technology* 6, 403–409.

Odell, J.T., Nagy, F. and Chua, N.-H. (1985) Identification of DNA sequences required for activity of the cauliflower mosaic virus 35S promoter. *Nature* 313, 810–812.

Osbourn, J.K., Watts, J.W., Beachy, R.N. and Wilson, T.M.A. (1989) Evidence that nucleocapsid disassembly and a later step in virus replication are inhibited in transgenic tobacco protoplasts expressing TMV coat protein. *Virology* 172, 370–373.

Palukaitis, P. and Zaitlin, M. (1984) A model to explain the 'cross-protection' phenomenon shown by plant viruses and viroids. In *Plant–Microbe Interaction: Molecular and Genetic Perspectives*, ed. T. Kosuge and E.W. Nester. Macmillan, New York, pp. 420–429.

Powell, P.A., Stark, D.M., Sanders, P.R. and Beachy, R.N. (1989) Protection against tobacco mosaic virus in transgenic plants that express TMV antisense RNA. *Proceedings of the National Academy of Sciences (USA)* 86, 6949–6952.

Powell, P.A., Sanders, P.R., Tumer, N., Fraley, R.T. and Beachy, R.N. (1990) Protection against tobacco mosaic virus infection in transgenic plants requires accumulation of coat protein rather than coat protein RNA sequences. *Virology* **175**, 124–130.

Powell-Abel, P., Nelson, R.S., De, B., Hoffman, N., Rogers, S.G., Fraley, R.T. and Beachy, R.N. (1986) Delay of disease development in transgenic plants that express the tobacco mosaic virus coat protein gene. *Science* **232**, 738–743.

Rast, A.T.B. (1972) MII-16, an artificial symptomless mutant of tobacco mosaic virus for seedling inoculation of tomato crops. *Netherlands Journal of Plant Pathology* **78**, 110–112.

Register, J.C. and Beachy, R.N. (1988) Resistance to TMV in transgenic plants results from interference with an early event in infection. *Virology* **166**, 524–532.

Register, J.C. and Beachy, R.N. (1989) Effect of protein aggregation state on coat protein-mediated protection against tobacco mosaic virus using a transient protoplast assay. *Virology* **173**, 656–663.

Ross, H. (1986) Potato breeding – problems and perspectives. *Advances in Plant Breeding*, No. 13. Paul Povey, Berlin, Hamburg, pp. 132.

Sanders, P.R., Winter, J.A., Barnason, A.R., Rogers, S.G. and Fraley, R.T. (1987) Comparison of Cauliflower mosaic virus 35S and nopaline synthase promoters in transgenic plants. *Nucleic Acids Research* **15**, 1543–1557.

Sanford, J.C. and Johnston, S.A. (1985) The concept of parasite-derived resistance genes from the parasites own genome. *Journal of Theoretical Biology* **113**, 395–405.

Sarver, N., Cantin, E.M., Chang, P.S., Zaia, J.A., Ladne, P.A., Stephens, D.A. and Rossi, J.J. (1990) Ribozymes as potential anti-HIV-1 therapeutic agents. *Science* **247**, 1222–1225.

Sequeira, L. (1984) Cross protection and induced resistance: their potential for plant disease control. *Trends in Biotechnology* **2**, 25–29.

Sheeny, R.E., Kramer, M. and Hiatt, W.R. (1988) Reduction of polygalacturonase activity in tomato fruit by antisense RNA. *Proceedings of the National Academy of Sciences (USA)* **85**, 8805–8809.

Sleat, D.E. and Palukaitis, P. (1990) Site-directed mutagenesis of a plant viral satellite RNA changes its phenotype from ameliorative to necrogenic. *Proceedings of the National Academy of Sciences (USA)* **87**, 2946–2950.

Smith, C.J.S., Watson, C.F., Ray J., Bird, C.R., Morris, P.C., Schuch, W. and Grierson, D. (1988) Antisense RNA inhibition of polygalacturonase gene expression in transgenic tomatoes. *Nature (London)* **334**, 724–726.

Stanley, J. and Townsend, R. (1985) Characterisation of DNA forms associated with cassava latent virus infection. *Nucleic Acids Research* **13**, 2189–2205.

Stanley, J., Frischmuth, R. and Ellwood, S. (1990) Defective viral DNA ameliorates symptoms of geminivirus infection in transgenic plants. *Proceedings of the National Academy of Sciences (USA)* **87**, 6291–6295.

Stark, D.M. and Beachy, R.N. (1989) Protection against potyvirus infection in transgenic plants: evidence for broad spectrum resistance. *Bio/Technology* **7**, 1257–1262.

Stark, D.M., Register, J.C., Nejidat, A. and Beachy, R.N. (1990) Toward a better understanding of coat protein mediated protection. In *Plant Gene Transfer*, eds. C.J. Lamb and R.N. Beachy. Alan R. Liss Inc., New York, pp. 275–287.

Takanami, Y. (1981) A striking change in symptoms on cucumber mosaic virus-infected tobacco plants induced by a satellite RNA. *Virology* **109**, 120–126.

Tien, P., Li, C.L., Jiang, Z.R., Zhao, S.Z., Yie, Y. and Liu, Y.Z. (1990b) Transgenic Virus resistance in transgenic plants that express the monomer gene of cucumber mosaic virus satellite RNA in greenhouse and field. *Abstracts of VIIIth International Congress of Virology*, Berlin, 24–26 August, 1990, p. 123.

Tien, P., Li, C.L., Jiang, Z.R., Zhao, S.Z., Yie, Y. and Liu, Y.Z. (1990b) Transgenic pepper plants resistant to cucumber mosaic virus expressing its satellite cDNA. *Abstracts of VIIIth International Congress of Virology*, Berlin, 24–26 August, 1990, p. 454.

Tumer, N.E., O'Connell, K.M., Nelson, R.S., Sanders, P.R., Beachy, R.N., Fraley, R.T. and Shah, D.M. (1987) Expression of alfalfa mosaic virus coat protein gene confers cross-protection in transgenic tobacco and tomato plants. *EMBO Journal* **6**, 1181–1188.

Tumer, N.E., Haley, L., Kaniewski, W. and Sanders, P. (1990) Analysis of the mechanism of resistance in transgenic tobacco expressing AlMV coat protein. *Abstracts of VIIIth International Congress of Virology*, p. 456.

van den Elzen, P.J.M., Huisman, M.J., Willink, D.P.-L., Jongedijk, E., Hoekema, A. and Cornelissen, B.J.C. (1989) Engineering virus resistance in agricultural crops. *Plant Molecular Biology* **13**, 337–346.

van der Krol, A.R., Lenting, P.E., Veenstra, J., van der Meer, I.M., Koes, R.E., Gerats, A.G.M., Mol, J.N.M. and Stuitje, A.R. (1988) An antisense chalcone synthase gene in transgenic plants inhibits flower pigmentation. *Nature (London)* **333**, 866–869.

van Dun, C.M.P. and Bol, J.F. (1988) Transgenic tobacco plants accumulating tobacco rattle virus coat protein resist infection with tobacco rattle virus and pea early browning virus. *Virology* **167**, 649–652.

van Dun, C.M.P., Bol, J.F. and van Vloten-Doting, L. (1987) Expression of alfalfa mosaic virus and tobacco rattle virus coat protein genes in transgenic tobacco plants. *Virology* **159**, 299–305.

van Dun, C.M.P., van Vloten-Doting, L. and Bol, J.F. (1988a) Expression of alfalfa mosaic virus cDNA1 and 2 in transgenic tobacco plants. *Virology* **163**, 572–578.

van Dun, C.M.P., Verduin, B., van Vloten-Doting, L. and Bol, J.F. (1988b) Transgenic tobacco expressing tobacco streak virus or mutated alfalfa mosaic virus coat protein does not cross-protect against alfalfa mosaic virus infection. *Virology* **164**, 383–389.

Wang, H.-L., Yeh, S.-D., Chiu, R.-J. and Gonsalves, D. (1987) Effectiveness of cross-protection by mild mutants of papaya ringspot virus for control of ringspot disease of papaya in Taiwan. *Plant Disease* **71**, 491–497.

Waterworth, H.E., Kaper, J.M. and Tousignant, M.E. (1979) CARNA 5, the small cucumber mosaic virus-dependent replicating RNA, regulates disease expression. *Science* **204**, 845–847.

Wisniewski, L.A., Powell, P.A., Nelson, R.S. and Beachy, R.N. (1990) Local and systemic spread of tobacco mosaic virus in transgenic tobacco. *Plant Cell* **2**, 559–567.

Wu, X., Beachy, R.N., Wilson, T.M.A. and Shaw, J.G. (1990) Inhibition of uncoating of tobacco mosaic virus particles in protoplasts from transgenic tobacco plants that express the viral coat protein gene. *Virology* **179**, 893–895.

Yamaya, J., Yoshioka, M., Meshi, T., Okada, Y. and Ohno, T. (1988) Cross protection in transgenic tobacco plants expressing a mild strain of tobacco mosaic virus. *Molecular and General Genetics* **215**, 173–175.

Yeh, S.-D. and Gonsalves, D. (1984) Evaluation of induced mutants of papaya ringspot virus for control by cross protection. *Phytopathology* **74**, 1086–1091.

Young, M. and Gerlach, W.L. (1990) Ribozyme activity against plant pathogen RNAs. *Abstracts of Workshop on Genome Expression and Pathogenesis of Plant RNA Viruses*. Serie Universitaria 253, Fundacion Juan March, Madrid, p. 31.

Zaitlin, M. (1976) Viral cross protection: more understanding is needed. *Phytopathology* **66**, 382–383.

Chapter 9
Potential of Secondary Metabolites in Genetic Engineering of Crops for Resistance

David L. Hallahan, John A. Pickett,
Lester J. Wadhams, Roger M. Wallsgrove
and Christine M. Woodcock

*AFRC Institute of Arable Crops Research,
Rothamsted Experimental Station, Harpenden,
Hertfordshire, AL5 2JQ, UK*

Introduction

'Secondary metabolite' is a loosely defined but convenient term for those organic compounds not directly involved in the primary metabolic processes of plants, such as photosynthesis, respiration or the biosynthesis of proteins. Secondary metabolites include a wide variety of compounds, some classified generally by chemical function, e.g. plant phenolics, and others, with widely varied functionality, by terms that relate to their biosyntheses, e.g. the isoprenoids (terpenoids). They are not necessarily produced by exclusive biosynthetic pathways and include compounds, such as non-protein amino acids and non-lipid fatty acids, that arise from pathways with steps common to primary metabolism. None the less, secondary metabolites are usually generated and stored away from primary functions and are often found in specialized structures such as foliar hairs (trichomes), or located in vacuoles. For a general review of the biochemistry of these compounds, see Goodwin and Mercer (1983).

1. Secondary metabolites in plant defence

1.1 Overview

Plant secondary metabolites contribute to defence against various organisms, including fungal pathogens, other plants and herbivorous animals ranging from insects to man. These metabolites may act as general toxicants or biocides, or may have a specific toxic action directed at a particular target. Alternatively, they may act as chemical messengers or signals with purely behavioural or regulatory effects. Such compounds are termed 'semiochemicals' and act as external hormones. Semiochemicals acting between members of the same species are terms 'pheromones' and, between different species, 'allomones' (where there is advantage to the emitting organism) or 'kairomones' (where there is advantage to the recipient organism). Some toxic secondary metabolites may be detected by the invading organism before there is a lethal effect, thereby acting as plant allomones. Alternatively, such allomones may have essentially no toxicity themselves, but simply convey the message that the plant should not be eaten. Secondary metabolites may also be involved in interactions disadvantageous to the plant which produces them. Thus, although some are toxic to certain attacking organisms, particularly those that have a general feeding or infective strategy, they may also be employed by other organisms that have evolved mechanisms for overcoming the toxic effects, as cues for feeding or development. Such host kairomones are common and also include compounds acting purely as semiochemicals, without themselves having obvious toxic effects. For general reviews of these aspects, see Shorey and McKelvey (1977) and Spencer (1988).

Biosynthesis of secondary metabolites in plants usually involves a number of enzymic steps and thereby a series of associated genes, thus presenting a difficult target for genetic engineering. This is aggravated if more than one biosynthetic pathway is required, e.g. in the biosynthesis of fatty acid esters of terpenoid alcohols. Furthermore, for appropriate storage and biological effects, the genes for the biosynthetic pathway may need to be expressed in particular tissues, and the genes for production of such tissue may need to be engineered into the transgenic plant for effective use of secondary metabolites against an invading organism. None the less, the possibilities for accomplishing such objectives exist and are actively being pursued (Dawson *et al.*, 1989a). One simplifying aspect is that almost all plants, and certainly all crop plants, have secondary metabolism, although it is somewhat limited in the cereals, a common feature of the Gramineae (= Poaceae). Thus, genetic engineering strategies may be devised that only require the limited modification of existing pathways. For example, providing that the genes for particular enzymes are appropriately expressed, they could be employed to augment existing pathways so as to enable biosynthesis of metabolites useful in crop plant protection.

The general benefits of secondary metabolism in the self-defence mechanisms of wild plants are undisputed. For crop plants, it can play a minor role and has often been sacrificed for improved yield and nutritional value, with protection provided by man, either culturally or by application of synthetic chemicals. None the less, there has been stark evidence for the deleterious effects of interfering with crop plant secondary metabolism. For example, the early experimental varieties of oilseed rape, *Brassica napus* (Cruciferae = Brassicaceae), which were produced with reduced glucosinolate content in the seed, were noticeably more susceptible to pest and disease damage. The reduction in glucosinolate content was necessary for herbivores, in this case cattle, which are not adapted to the chemical defences of crucifers, to accept the seed residue after oil extraction. However, the associated interference with the secondary metabolism responsible for defending the vegetative parts of the plant allowed increased damage by diseases and pests, particularly by unadapted organisms such as slugs. Such observations are important in assessing the likely value of secondary metabolites as targets for genetic modification of crop plants. The particular case of glucosinolate-based defence in crucifers will be discussed in more detail later. The enjoyable flavour of many crop products derives from secondary metabolites, such as those arising from the glucosinolates in the leaves of cabbages and Brussels sprouts (*Brassica oleracea* var.). However, there are dangers in modifying secondary metabolism; for example, the defence pathways in potatoes, *Solanum tuberosum* (Solanaceae), and celery, *Apium graveolens* (Umbelliferae = Apiaceae), which produce toxic alkaloids and photodermatitic agents respectively, have been incorporated into the edible parts by plant breeding programmes attempting to develop pest-resistant cultivars.

Almost all chemicals, whether natural or synthetic, are toxic at some dose level, but, by concentrating on targets involving semiochemicals acting by non-toxic modes of action, safe approaches to the use of secondary metabolites can be developed. In addition, tissue-specific promoter gene sequences will allow expression of new pathways within the plant, away from the edible parts of the crop. By adopting a strategic approach to modifying secondary metabolism, the particular compounds involved can be defined, allowing the accurate assessment of risks which would be required for registration purposes.

The ideas put forward in this overview will be developed for specific examples. The secondary metabolites of potential value in crop protection will be presented according to their general mode of action, and the possibilities for development by genetic engineering will be considered. The current status of enzyme isolation and genetic studies on secondary metabolism will be discussed and current progress and further needs in related plant molecular genetics will be reviewed.

1.2 Toxicants

Biocides

A common form of defence by plants is the production of broadly toxic secondary metabolites or biocides that give protection against a wide range of attacking or competing organisms. Plant phenolics can fulfil such a role (Harborne, 1985). The phenolic acids and polyphenols, such as the tannins, show a very broad spectrum of activity. They can function as uncouplers of oxidative phosphorylation or more generally by reacting chemically with proteins, thereby destroying essential enzymic activity, for example in the gut of insect herbivores. The compounds may be only weakly active and so are present in large amounts. Their phytotoxicity is often accommodated by storage as glycosides. Defence against micro-organisms can be provided by phenolic phytoalexins, which are biosynthesized and accumulated in the tissues after initial damage (Friend, 1985). There are, of course, other chemically distinct groups of phytoalexins (Bailey, 1982; Deverall, 1982), but the strategy of producing the toxicant on damage is the same. The phenolic phytoalexins may also display high toxicity against herbivores, but some structure/activity relationships may be discerned between effects on animals and antimicrobial action. Thus, in the case of the furanocoumarins, bergapten (I) is found as a phytoalexin in celery, but the related compound xanthotoxin (II), from parsnips, *Pastinaca sativa* (Umbelliferae), is a phototoxin against some herbivores (Harborne, 1985; Zobel and Brown, 1990). These compounds provide examples of photoactivation, another strategy adopted by plants to generate biocidal compounds from more easily stored precursors (Downum, 1986).

Alkaloids are a notable group of secondary metabolites with general toxicity to herbivores, although of less importance in disease resistance. They can be involved in highly developed adaptations by herbivores specializing on the producer plant. For example, the common ragwort, *Senecio jacobaea* (Compositae = Asteraceae), has developed a general protection against herbivores by producing pyrrolizidine alkaloids such as senecionine (III). These compounds appear to be synthesized as the *N*-

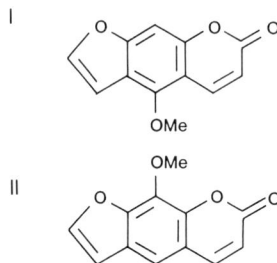

oxides in the roots and are then translocated in the phloem to accumulate as the alkaloids in the shoots (Hartmann *et al.*, 1989). However, the cinnabar moth, *Tyria jacobaeae*, by feeding on the leaves, sequesters considerable amounts of these toxic alkaloids for its own protection, which it advertises to predators by its bright aposematic coloration. Plant alkaloids can also act as a starting material for synthesis of novel alkaloids by adapted insects (Hartmann *et al.*, 1991).

As well as the phenolics and alkaloids, there are a great many general toxophores stored or transported within plants as glycosides. Glycosylation also provides the means of utilizing highly volatile or unstable compounds such as hydrogen cyanide and organic isothiocyanates in pest resistance. Hydrogen cyanide is often stored as cyanogenic glycosides, which are common in the plant kingdom, for example in protecting the seed of species in the Rosaceae (Nahrstedt, 1987). The hydrogen cyanide is released by the sequential action of a β-glycosidase and an oxynitrilase, which come into contact with the glycoside substrate on insect damage. The catabolism of dhurrin (IV), a simple cyanogenic glycoside from *Sorghum vulgare* (Gramineae), is given in Scheme 1.

Plants, particularly those in the Cruciferae, employ glucosinolates (V), which release organic isothiocyanates (VI) by the action of thioglucosidase (myrosinase) enzymes (Scheme 2). The isothiocyanates act potently against many herbivores and pathogens not adapted to deal with such toxic compounds (Vaughan *et al.*, 1976; Fenwick *et al.*, 1983).

The plant kingdom also contains a large variety of compounds which are cytotoxic, often through being acceptors of nucleophiles in Michael addition reactions. These compounds can be potentially valuable leads for anticancer drugs, but may be biocidal and toxic to man. A common group are the sesquiterpene lactones, which fall into four main classes, the

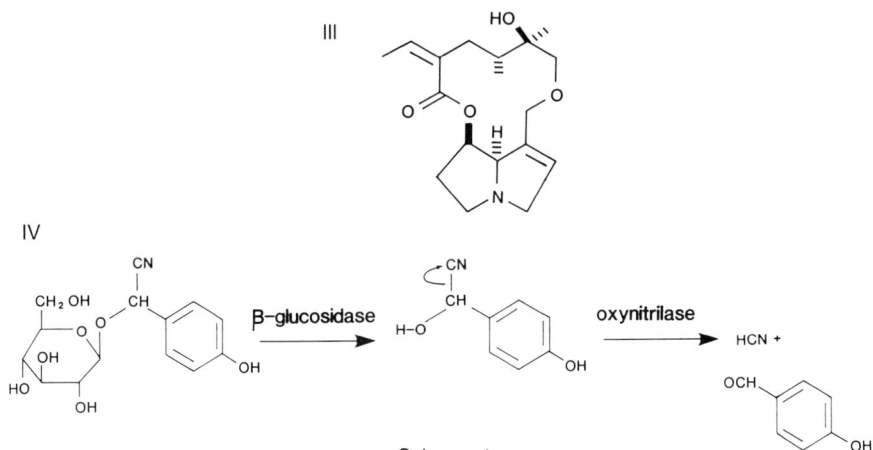

Scheme 1

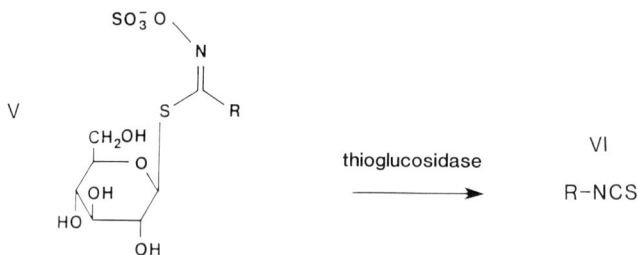

$SO_3^- O$

N

V

S R

CH_2OH

O

OH

HO

OH

thioglucosidase

⟶

VI

R-NCS

Scheme 2

germacrolides, eudesmanolides, heliangolides and guaianolides, depending on their sesquiterpene skeleton e.g. guaianolide (VII). A potent example is 8 β-sarracinoyloxycumambranolide (VIII) from *Helianthus maximilani* (Compositae) (Gershenzon *et al.*, 1985). Various functionalities are found in these compounds, but the exomethylene group in conjugation with the lactone carbonyl group is a typical feature and is a site for Michael addition reactions, presumably involving protein nucleophiles such as the cysteine-SH and lysine-NH_2 groups.

Saponins are another class of generally toxic compounds from plants. They disrupt biological membranes by means of surfactant properties arising from a steroidal, spirostanol or triterpenoidal hydrophobic moiety, with the hydrophilic part usually provided by 3-*O*-polyglycosylation (Hiller, 1987).

Specific toxicants

Many groups of general biocides have members that show selective activity. For example, certain terpenoid saponins, e.g. oleanoglycatoxin-A (IX) from *Phytolacca dodecandra* (Phytolaccaceae), have very high molluscicidal activity (Domon and Hostettman, 1984) and are being developed as natural mollusc control agents (Marston and Hostettman, 1991). A common strategy in producing specific agents by synthesis is to modify the structure of a biocide so that it is inactive until it reaches a particular target within the pest. An alternative is to generate the biocide as a non-toxic precursor that is then converted into a toxic agent by a process specific to the target organism, and this approach has been exploited in the production of synthetic pesticides as propesticides (Prestwich, 1990). Other more generally toxic agents may show specificity through differences in detoxifying ability between organisms. Thus, pyrethrin I (X), a highly insecticidal compound from the pyrethrum daisy, *Chrysanthemum* (= *Tanacetum*) *cinerariifolium* (Compositae), shows its valuable selectivity only because other organisms and particularly mammals, including man, are rapidly able to detoxify such compounds by mixed-function oxidase activity before they attack the nervous system (Casida, 1973).

VII

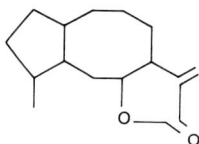

VIII

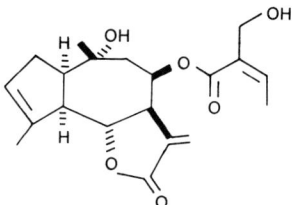

IX

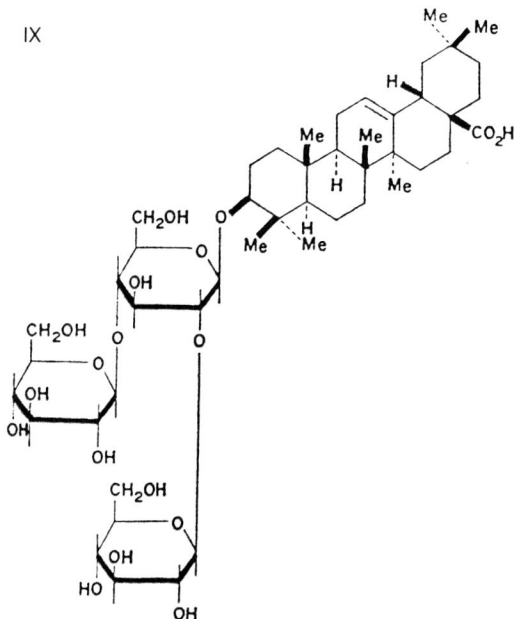

X

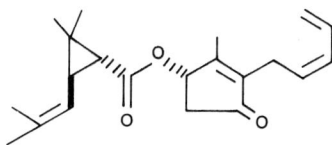

Some plant toxins are active only against insects or arthropods because they act as agonists of specific hormones (Bowers, 1985) or otherwise interfere with hormonal actions. The juvenile hormones (JHs) are responsible for regulating a number of processes in insects, most notably in maintaining the larval state until pupation is required. The natural JHs are simple, uncyclized sesquiterpenoids or homosesquiterpenoids. Some plants produce particularly potent analogues and the grasshopper cyperus, *Cyperus iria*, a Malaysian plant in the Cyperaceae, actually produces JH III (XI), which causes marked morphogenic effects in insects (Toong *et al.*, 1988). Many related sesquiterpenoids have some activity agonistic to the JHs (Mauchamp and Pickett, 1987). Other plants contain compounds active as anti-JHs, e.g. the precocenes such as precocene II (XII) from *Ageratum houstonianum* (Compositae) (Bowers, 1991).

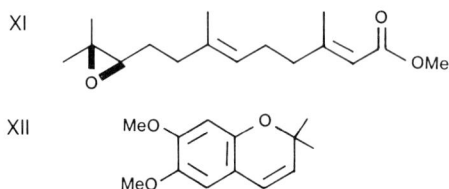

Ecdysis (moulting) in insects and other arthropods is controlled by the ecdysteroid hormones, and many plants produce phytoecdysteroids, which are toxic by interfering with this process. β-Ecdysone (XIII) is very common in members of the Chenopodiaceae, such as fat hen, *Chenopodium alba*, and is generally insecticidal (Báthori *et al.*, 1987). However, some phytoecdysteroids have more specific and potent activity, for example 29–norsengosterone (XIV) from *Ajuga remota* (Labiatae = Lamiaceae), which is very effective against whitefly (Aleyrodidae) (Camps, 1991).

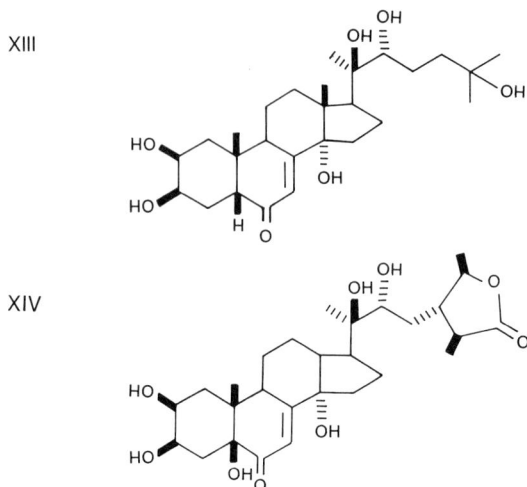

Many more classes of biocidal or specifically toxic agents can be found in plants which reduce herbivory and development of pathogens (Friend and Threlfall, 1976; Green and Hedin, 1986). Often, plants also produce allelopathic agents that are phytotoxic and thereby decrease competition from other plants (Waller, 1987). However, these phytotoxic agents are generally weak biocides, produced in very large amounts.

Exploitation of toxicants

Plant-derived toxicants have long been exploited in crop protection. The extract of the pyrethrum daisy containing pyrethrin I, mentioned above, is widely used indoors to control insects and provided the lead compound in the development of the pyrethroids, a highly successful class of synthetic insecticides (Elliott, 1990). However, there are few examples of natural products leading to such successful pesticides (Elliott, 1979). Pyrethrin I represented a particularly good lead in that it is highly toxic to insects and has low mammalian toxicity, and had only one disadvantage in being too unstable for field use. Natural products currently under investigation as leads for pesticides include azadirachtin (XV) from the Indian neem tree, *Azadirachta indica* (Meliaceae) (Kraus *et al.*, 1985; Broughton *et al.*, 1986). This compound is typical of most potential natural product leads in that it is probably too unstable and insufficiently active as an insecticide itself, but has too complicated a structure to allow enough analogues to be produced to overcome these problems. None the less, some very innovative synthetic approaches are being made (Ley, 1990).

Genetic manipulation could be used to increase the yields of valuable toxicants from plants, although this approach would be extremely difficult with compounds such as pyrethrin I, which has a complicated biosynthesis involving both isoprenoid and polyketide pathways, or azadirachtin, which has many more oxidative steps even after construction of the complicated tetranortriterpenoid skeleton. However, particular steps, difficult to accomplish chemically, could be achieved in a fermentor, using genetically modified micro-organisms.

For genetic modification of crop plants to produce toxicants of value in pest control, there are two possible strategies. One is to modify existing biocide production to improve performance. Man has a well-developed detoxification system which can deal with relatively high levels of toxicants in food, e.g. the cinnamic acid derivatives in coffee with known rodent carcinogenicity (Ames and Gold, 1990). Also, some toxic materials, such as the cyanogenic glycosides which protect the roots of cassava, *Manihot esculenta* (Euphorbiaceae), during growth, can be removed by washing etc. before consumption. However, with plant genetic modification, it is possible that production of biocides could be confined to parts of the plant not eaten. Thus, to produce pest-resistant potatoes, the objective would be to express genes for the biosynthetic pathways of the toxic glycoalkaloids found in wild-type cultivars, and other *Solanum* species, in as much of the plant as possible, but not in the tubers prior to harvest. In addition, if damage-induced signals were to be used, perhaps with the signal applied exogenously to the crop, then genes for biosynthesis of the defence compounds could be expressed only when pest monitoring suggested a need. Tissue-specific expression and wound-inducible signals are discussed more fully in the final section of this chapter.

As mentioned earlier, the biocidal isothiocyanate catabolites of gluco-sinolates found in oilseed rape are not wanted in the seed, but initial attempts to interfere with glucosinolate production left the vegetative parts of the plant very vulnerable to many pathogens and unadapted herbivores. Even the modern low seed glucosinolate cultivars, which have much better pest resistance through relatively high vegetative glucosinolate levels, can be more readily predated by unadapted herbivores, such as slugs (Glen *et al.*, 1989), pollen beetles (*Meligethes* spp.) (Milford *et al.*, 1989) and the peach–potato aphid *Myzus persicae* (A.J.R. Porter, unpublished), and also by some fungal pathogens (Rawlinson *et al.*, 1989). The objective now is to identify the R groups in the glucosinolates (Scheme 2) that are most valuable in reducing disease and damage by unadapted pests, and then to control the expression so that these are produced in the vegetative parts, but with low glucosinolate levels in the seed. A further objective is to reduce vegetative expression of compounds where R enhances develop-ment and colonization by adapted pests (Dawson *et al.*, 1989a).

The other strategy for exploiting toxicants in plants is to add alien genes for enzymes to augment existing biosynthetic pathways so as to produce secondary metabolites more active in pest resistance than the endogenous components. Thus, if the appropriate genes from *Ajuga remota* could be cloned, transferred and suitably expressed in spinach, *Spinacia oleracea* (Chenopodiaceae), then, instead of producing the weakly active phytoecdysteroid β-ecdysone, a more active analogue such as 29-norsengosterone (XIV) could be produced.

Many other examples may be cited, but much pioneering work needs to be done on carefully selected cases. It may be more advisable, since such work is for the long-term future, to consider plant compounds active by non-toxic mechanisms, e.g. the semiochemicals (see below). However, as reported recently (Ames *et al.*, 1990), 99.99% of pesticides found in the diet are natural plant components and it should therefore not be difficult to register transgenic plants protected by natural toxicants provided that these are well defined, are produced by rational genetic modifications and can be shown to be absent from the product to be consumed.

1.3 Semiochemicals

Host location and selection by phytophagous insects is dependent on an array of stimuli associated with both host and non-host plants, involving visual, olfactory, mechanical and gustatory signals. Whilst the first two cues are usually associated with long-range host finding, host selection or rejection frequently depends on contact chemoreception of stimulants or deterrents.

Host location: volatile semiochemicals

Volatile signals from plants are utilized by phytophagous insects in host plant location and can be broadly divided into two groups: those that occur commonly and are derived from biosynthetic pathways present throughout the plant kingdom, and the more restricted chemicals that are found in only a few plant species.

Green leaf volatiles, which comprise a group of six-carbon alcohols, aldehydes and ester derivatives produced from unsaturated fatty acids, are responsible for the characteristic odour of damaged leaves, e.g. freshly mown grass. They have been implicated in mediating certain insect–plant interactions and have also been shown to synergize the activity of a number of coleopteran (beetle) aggregation pheromones (Blight *et al.*, 1984; Dickens, 1989; Dickens *et al.*, 1990). The green leaf volatiles form an essential component of the odour which attracts the Colorado potato beetle, *Leptinotarsa decemlineata*, to its host (Visser and Avé, 1978). However, wind tunnel studies demonstrated that, although none of these compounds was active alone, when added to potato leaf odour they prevented the beetles from locating the plants. Thus, attraction of the Colorado beetle to its host appears to be dependent on the ratios of the green leaf volatiles in the plant odour. Genetic modification of the plant's secondary metabolism to alter these ratios could therefore provide a simple method of interfering with host location by this pest.

A large number of phytophagous insects possess olfactory receptors for green leaf volatiles and many show behavioural responses to them (Visser, 1983, 1986). These compounds have been implicated in interactions at the third trophic level. Parasitic wasps use a variety of host-produced kairomones in host-searching, including the volatile signals released by the plant on which the host is feeding. Female *Microplitis croceipes*, a braconid parasitoid of the corn earworm (*Heliothis zea*), and the ichneumonid parasite, *Netelia heroica*, respond to green leaf volatiles in the wind tunnel (Whitman and Eller, 1990). However, the two species show different responses to the individual compounds, which may reflect their specialization on particular plant taxa. Modifying the content and ratios of the volatiles produced by crop plants could allow a wider range of habitats to be searched by useful parasites and parasitoids.

In addition to the activity of general plant volatiles, highly specific compounds found in only a few species also elicit behavioural activity. Host-specific phenylpropanoids are involved in the attraction of the carrot fly, *Psila rosae*, to its host plant (Guerin *et al.*, 1983). Similarly, the volatile isothiocyanate catabolites (VI) of glucosinolates are implicated in host selection by cruciferous pests. Electrophysiological studies on aphid antennae found olfactory receptors for these compounds (VI, R = alkenyl), and laboratory olfactometer studies demonstrated that crucifer specialists such as the mealy cabbage aphid *Brevicoryne brassicae*, and the turnip aphid, *Lipaphis erysimi*, are attracted by them (Nottingham *et al.*, 1991). For *L. erysimi*, the isothiocyanates have the additional function of synergizing the activity of the alarm pheromone, (E)-β-farnesene (Dawson *et al.*, 1987), which is produced by the aphids when they are attacked. Since it is likely that the isothiocyanates arise from the host plant and that feeding aphids play an active role in their production, the elimination of the allyl, 3-butenyl and 4-pentenyl glucosinolates (sinigrin, gluconapin and glucobrassicanapin respectively) would be highly beneficial, not only in reducing host plant location by this aphid, but also by interfering with its ability to defend itself against predation.

Behavioural work on other crucifer specialists has largely concentrated on the effects of allyl isothiocyanate. Recent electrophysiological studies on two coleopteran pests of oilseed rape, *Ceutorhynchus assimilis* and *Psylliodes chrysocephala*, have shown that these insects have few receptors for this compound, and that perception of isothiocyanates is mediated through a mixture of generalist and specialist olfactory cells. For *C. assimilis*, the specialist cells clearly discriminate between aromatic and alkenyl isothiocyanates (Blight *et al.*, 1989) and, although the behavioural role of these compounds has yet to be determined, it is likely that the elimination of alkenyl glucosinolates in the crop plant would again prove beneficial in reducing pest pressure.

Aphids not adapted to crucifers, such as the black bean aphid, *Aphis fabae*, and the damson–hop aphid *Phorodon humuli*, also possess receptors for the alkenyl isothiocyanates, and laboratory studies have shown that they are repelled by these compounds. Many traditional cultivation methods of reducing pest damage involve disrupting host location by intercropping with non-host plants (Matthews *et al.*, 1983). In the laboratory bioassay, 4–pentenyl isothiocyanate inhibited the attraction of *A. fabae* to its host plant (Nottingham *et al.*, 1991) and field application of a chemical slow-release formulation to hops decreased colonization of the crop by *P. humuli* spring migrants (Pickett *et al.*, 1991). Thus, it is now possible to demonstrate, at the molecular level, the basis of mixed cropping techniques and to provide targets for selective modification of crop plants to reduce pest colonization.

Host selection: antifeedants/phagostimulants

Many plants produce non-volatile allomones that interact with insects' gustatory receptors and prevent feeding by generalist herbivores (Jacobson, 1990). Some of these, in addition to insecticidal activity (see above), can act as antifeedants, e.g. azadirachtin (XV), which is effective against lepidopteran larvae down to a few parts per million in laboratory assays. Antifeedants which are also highly active but have less complicated structures are the drimane sesquiterpenoids such as warburganal (XVI) from *Warburgia ugandensis* (Canellaceae) and polygodial (XVII) from the water-pepper, *Polygonum hydropiper* (Polygonaceae). In the laboratory, application of (−)-polygodial reduced colonization by aphids, including those highly resistant to conventional insecticides, and field application to cereals decreased transmission of barley yellow dwarf virus by the bird-cherry-oat aphid, *Rhopalosiphum padi*, giving over a tonne per hectare increase in yield relative to untreated plots (Pickett *et al.*, 1987). The synthetic compound 9-α-hydroxydrimenal (XVIII), although inactive against aphid colonization, is highly active against lepidopteran larvae (Asakawa *et al.*, 1988), and its precursor, drimenol (XIX), is produced in relatively high yields by certain Basidiomycetes (Hanssen and Abraham, 1987). The biosynthesis of (−)-polygodial and related drimanes is discussed in the next section of this chapter. These plant-derived allomones have a broad spectrum of activity and show considerable promise in the management of insect pests, either by direct application to crops or by incorporation into the crop plants themselves.

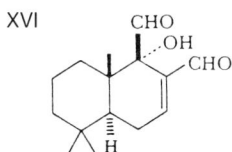

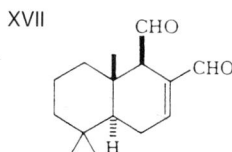

XVIII

HO CHO

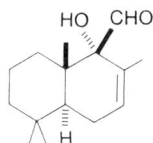

XIX

CH₂OH

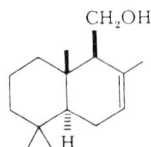

As discussed earlier, many plants have developed toxicants to provide protection against generalist herbivores, but some insects have become specialized to feed on these plants and utilize the compounds as feeding stimulants. Thus, the intensely bitter and toxic cucurbitacins, e.g. XX, produced by members of the Cucurbitaceae, are feeding stimulants for beetles in the genus *Diabrotica* (Metcalf *et al.*, 1980). The bitter components can be selectively accumulated by the beetles and deter feeding by some predators (Nishida and Fukami, 1990). Larvae of the turnip sawfly, *Athania rosae*, feed exclusively on plants in the Cruciferae. However, the adults feed voraciously on a verbenaceous plant, *Clerodendron trichotomum*, which contains feeding stimulants, e.g. clerodendrin D (XXI) (Nishida *et al.*, 1989). Sequestration of this compound in the body tissues provides some protection from predation.

Host selection and rejection of Cruciferae by insects is mediated by the glucosinolates which, although representing an important defence against unadapted pests and diseases, are utilized by a number of organisms as recognition signals of suitable host plants. Thus, while sinigrin is a deterrent for aphids not adapted to crucifers, it is a feeding stimulant for *Brevicoryne brassicae* (Wensler, 1962) and *Lipaphis erysimi* (Nault and Styer, 1972). Attempts to modify secondary metabolite production in crop plants must therefore be accompanied by a detailed knowledge of the roles of the compounds involved.

Pheromones related to plant secondary metabolites

Some pheromones utilized by insect pests are closely related or even identical to plant secondary metabolites. Of the pheromones, the best known are the lepidopteran sex pheromones which are used by females to attract males. Sex pheromones are already employed in traps for monitoring pest population levels to determine whether and when pesticide applications are required, e.g. for the pea moth, *Cydia nigricana* (Macaulay *et al.*, 1985). In some cases, it has been possible to use the pheromone to control the pests directly. A slow-release formulation of the sex pheromone of the pink bollworm moth, *Pectinophora gossypiella*, is used in the field to confuse males and disrupt normal mate location (Critchley *et al.*, 1985). An alternative strategy is to use the crop plant itself as a source of the sex

XX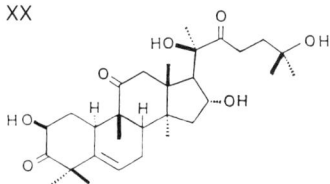

XXI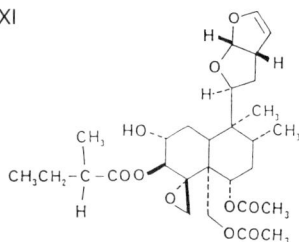

pheromone. The biosynthetic pathways to a number of these pheromones have been elucidated and typically involve dehydrogenation of long-chain fatty-acid acyl groups followed, where necessary, by shortening of the carbon chain (Scheme 3) (Bjostad and Roelofs, 1983). Reduction to the aldehyde or alcohol, followed by acetylation, yields the required pheromone. In the biosynthesis of (*Z*)-11-hexadecenal, a major component of the sex pheromones of the diamondback moth, *Plutella xylostella*, and the cotton bollworm, *Helicoverpa* (= *Heliothis*) *armigera*, a Δ^{11}-desaturase enzyme converts a palmitoyl group into the corresponding (*Z*)-11-hexadecenoyl group, giving the pheromone component on reduction. Certain plants, e.g. *Ophrys* spp. (Borg-Karlson, 1987; Borg-Karlson *et al.*, 1987), *Rosa rugosa* (Dobson *et al.*, 1987), *Mikania amara* (Da Silva *et al.*, 1984), *Rhus typhina* (Bestmann *et al.*, 1988) and a forage sorghum (Kami, 1975), are known to produce the saturated compounds, presumably from the palmitoyl precursor. The transfer of the Δ^{11}-desaturase enzyme into these plants could complete the biosynthetic pathway to (*Z*)-11-hexadecenal.

In temperate climates, aphids are the main insect pests and the sex pheromones of a number of aphids have recently been identified as specific isomers of the iridoid monoterpenes nepetalactone and nepetalactol (Dawson *et al.*, 1990). The sex pheromone of the hop aphid, *Phorodon humuli*, comprises the nepetalactols XXII and XXIII (Campbell et al., 1990). In the field, the pheromone catches large numbers of males and offers excellent potential for the control of this highly resistant pest. The nepetalactone precursor with the correct stereochemistry can be obtained in high yield from *Nepeta mussinii* (Labiatae). In a similar strategy to that envisaged for Lepidoptera, modification of the secondary metabolism of the hop, *Humulus lupulus* (Cannabaceae), to produce the aphid sex pheromone (see next section) could disrupt normal mate location by the males.

(*E*)-*β*-Farnesene, the main alarm pheromone component for many species of aphids, occurs commonly in plants, but does not generally interfere with aphid colonization, since biosynthesis in the plant from farnesyl pyrophosphate is accompanied by other sesquiterpene hydrocarbon cyclization products including (−)-*β*-caryophyllene, a potent inhibitor of the alarm pheromone activity (Dawson *et al.*, 1984). However,

Acyl precursor

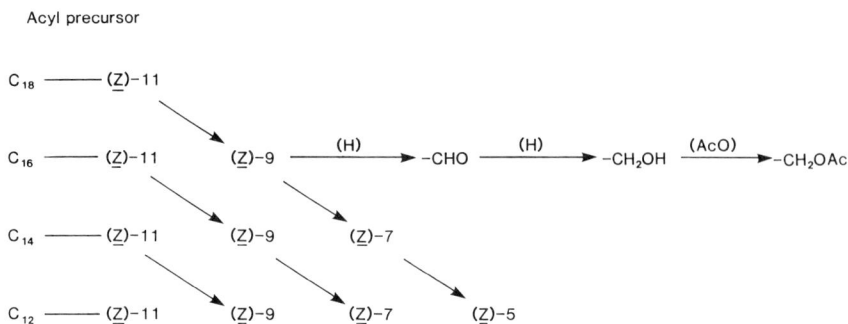

Scheme 3

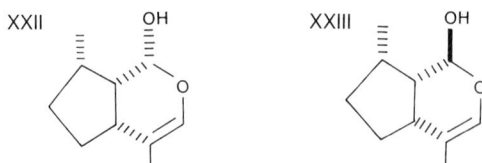

the Bolivian wild potato, *Solanum berthaultii,* a close relative of the commercial potato, produces (E)-β-farnesene from glandular trichomes on its leaves in a manner that imitates production of alarm pheromone by aphids, thereby deterring colonization. Although this would appear to be an attractive property for incorporation into crop plants, production of such specialized structures in the leaf would be a very difficult target for genetic manipulation studies.

With a knowledge of the identity and role of biologically active compounds and an understanding of the enzymic steps leading to their production, it is possible to envisage the transfer of genes controlling metabolic pathways which could either prevent production of a key component in an attractant blend, or promote the biosynthesis of an additional secondary component that would decrease colonization of the crop by the pest. A number of such approaches are discussed in the next section. However, for these strategies to be effective, a full understanding of the relevant chemical ecology is essential, besides overcoming many further obstacles to such plant genetic engineering.

2. Biosynthetic pathways for crop protection compounds

Rather than attempting to present a general review of plant secondary metabolite biosynthesis, examples have been chosen from specific aspects mentioned previously in this chapter that illustrate the possibilities and the problems of the associated gene manipulation. The first class of compounds to be described consists of a large and complex family, with an equally complex biosynthetic pathway. In this case, the objective is to maximize the value in crop protection, and the potential for transformation into other species is somewhat restricted. Two other classes of compounds will be discussed that have potential for transformation into host species that do not make closely related compounds, and where the crop protection mechanisms are quite different.

2.1. Glucosinolate biosynthesis

In excess of 80 different glucosinolates (V) have been identified in plant tissues. The R group derives from an amino acid in each case and in the *Brassica* species there are glucosinolates derived from methionine (aliphatic and alkenyl glucosinolates), phenylalanine (aromatic glucosinolates) and tryptophan (indolyl glucosinolates). The biosynthetic pathway can be conveniently divided into three phases. The first involves chain extension of the amino acid. No glucosinolates derived from methionine itself occur in oilseed rape, the major aliphatic compounds coming from methionine analogues with two or three extra carbon atoms. Similarly, the major aromatic glucosinolates in rape are produced from homophenylalanine, i.e. with an additional carbon atom. This chain extension involves transamination, acetylation of the resulting oxoacid, decarboxylation and then transamination back to an amino acid. Methionine:glyoxylate aminotransferases that are presumed to be involved in this pathway have been purified and characterized (Chapple *et al.*, 1990).

The second stage is the conversion of the amino acid into its corresponding glucosinolate (Scheme 4). Feeding studies have established that the pathway proceeds via the *N*-hydroxyamino acid and aldoxime. The

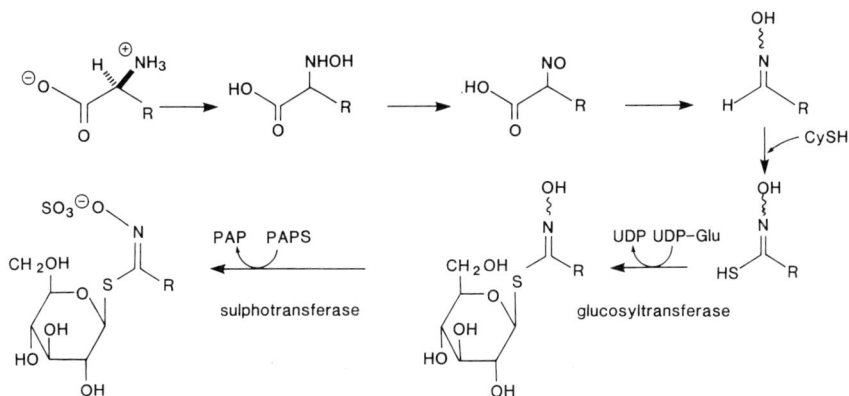

Scheme 4

sequence amino acid – hydroxyamino acid – aldoxime is common to some
other biosynthetic pathways, in particular that of cyanogenic glycosides
(McFarlane et al., 1975) and possibly the plant hormone indole-3-acetic
acid. The general role of aldoximes in plants has been reviewed
(Mahadevan, 1973). Metabolism of indole-3-acetaldoxime by plants that
do not make glucosinolates has been demonstrated (see Helmlinger et al.,
1985), suggesting that biosynthesis of aldoximes may be of general, or at
least widespread, occurrence in plants.

In *Sorghum bicolor* seedlings, a microsomal, membrane-bound
enzyme preparation has been described that converts tyrosine into its
aldoxime, in the presence of NADPH (McFarlane et al., 1975). This
aldoxime is a precursor to the cyanogenic glycoside, dhurrin (IV). It has
recently been shown that similar microsomal preparations from leaves of
Brassica napus catalyse the NADPH-dependent decarboxylation of homophenylalanine (R.M. Wallsgrove, unpublished), and the system has some
characteristics of a cytochrome P450-linked enzyme. It has been suggested
that the sorghum system does not involve a cytochrome P450 in this step
(Halkier et al., 1989), although a P450-linked enzyme catalyses a later step
in dhurrin synthesis. An apparently quite different enzyme has been
isolated from *Brassica campestris* that oxidizes tryptophan to indole-3-
acetaldoxime. This is a peroxidase that requires Mn^{2+} for activity (Ludwig-
Muller and Hilgenberg, 1988). Further characterization revealed this
plasma membrane-bound enzyme to be one of a group of peroxidase
isozymes with high pI (8.1–8.4), immunologically related to tobacco
peroxidases (Ludwig-Muller et al., 1990).

The conversion of amino acid into oxime occurs in two steps, hydroxyphenylalanine being a better precursor of benzylglucosinolate than phenylalanine in *Sinapis alba* (Cruciferae) and other species. Cell-free extracts
that converted hydroxyphenylalanine into phenylacetaldehyde oxime did not
require co-factors, and oxygen appeared to be the physiological hydrogen
acceptor (Kindl and Underwood, 1968). The oxime-producing preparations described above thus appear to consist of at least two enzymes. The
specificity of amino acid entry into the glucosinolate pathway is probably a
function of one of these enzymes, and it is possible that isozymes having
different specificity occur, with differential regulation, as suggested by
glucosinolate induction studies (Koritsas et al., 1989; Birch et al., 1990;
Doughty et al., 1991). In these studies, it was noted that 'induction' in
response to herbivory or infection did not apply to all glucosinolates, but
that some were preferentially synthesized. Indeed, infected young leaves
had a two-stage response, a rapid and transient synthesis of alkenyl
glycosinolates, followed by a slower but more sustained increase in
aromatic and indolyl glucosinolates (Doughty et al., 1991). This is
particularly interesting in view of the ability of certain coleopteran pests of
rape to discriminate between the different volatile glucosinolate metabo-

lites (Blight *et al.*, 1989). The changes in leaf glucosinolate content may reflect differential 'induction' of isozymes, or the availability of the parent amino acids. A detailed understanding of this part of the biosynthetic pathway will thus be essential for modifying the synthesis of selected glucosinolates, or whole classes such as the indolyls.

The final steps in the synthesis of the glucosinolate molecule are catalysed by a glycosylase and a sulphotransferase, and considerable progress has been made in purifying and characterizing these enzymes (Glendening and Poulton, 1990; Jain *et al.*, 1990). The wide specificity of these enzymes makes them of less interest with regard to modifying the pathway, but either might be a suitable target for antisense RNA to block the synthesis of glucosinolates within a specific tissue.

The third phase of glucosinolate biosynthesis involves the modification of the side-chain of intact glucosinolates. Progoitrin (V, R = 2-hydroxybutenyl), for example, is produced by hydroxylation of butenyl-glucosinolate (Rossiter *et al.*, 1990), itself derived from methylthiobutyl-glucosinolate. With the exception of the butenylglucosinolate hydroxylase, which appears to be another cytochrome P450 enzyme (D.C. James and J.T. Rossiter, unpublished), the enzymes involved in glucosinolate modification have not been studied. Identification of these enzymes is crucial for any strategy to increase, or block, synthesis of a specific glucosinolate. Blocking progoitrin synthesis would be a particularly useful tactic, as this is the most toxic, to mammals, of the glucosinolates commonly found in rapeseed meal.

Whilst transfer of the biosynthetic enzymes to species that do not naturally make glucosinolates might seem an excessive task, the apparent ubiquity of aldoxime synthesis in plants does reduce the problem. The genes for only three enzymes would need to be inserted to lead to indolylglucosinolate biosynthesis, for example, in any species that synthesizes indoleacetaldoxime. One other enzyme would of course have to be inserted to produce a crop protection system, namely myrosinase (see Scheme 2). Intracellular targeting of the inserted proteins would be required to separate glucosinolate from myrosinase in the intact cell. In glucosinolate-containing species, glucosinolates are stored in the vacuole, whilst myrosinase is a cytosolic enzyme (Luthy and Matile, 1984).

A more promising approach would be the modification of glucosinolate biosynthesis in oilseed rape and other agronomically important crucifers. Introduction of specific genes from other species, such as *Arabidopsis thaliana* or wild crucifers, could introduce novel glucosinolates into cultivated species, or boost synthesis of otherwise minor components. Coupled with a deletion of other glucosinolates, by antisense RNA technology or mutant selection, the glucosinolate spectrum of a crop could be dramatically altered to enhance crop protection, deter specialist feeders and reduce unwanted toxicity in harvested parts.

2.2 Polygodial and related terpenoid antifeedants

Drimane-type sesquiterpene antifeedants such as (−)-polygodial (XVII)
and (−)-warburganal (XVI) are assumed to be synthesized via a cyclization
of farnesyl pyrophosphate to drimenol, followed by several oxidation steps
to produce the final product. In *Polygonum hydropiper*, (−)-polygodial
synthesis in the shoot apparently occurs only during new leaf growth, prior
to leaf expansion (Dawson *et al.*, 1989a). Various attempts to produce
shoot, callus or suspension cultures that make polygodial have been only
partially successful (Banthorpe *et al.*, 1989), and this limitation in plant
material has made isolation and extraction of the putative biosynthetic
enzymes very difficult.

There has been a report of the extraction of a 'drimenol-producing'
cyclase from *P. hydropiper* tissues converting farnesyl pyrophosphate into
drimenol (Banthorpe *et al.*, 1989), but a more critical analysis of the
reaction products suggested that drimenol was not produced in detectable
quantities by any cell-free extracts (D.L. Hallahan, unpublished). Further
studies are in progress to identify the cyclase, using the wood-rot fungus
Gloeophyllum odoratum (Basidiomycetae), which synthesizes drimenol
(Hanssen and Abraham, 1987).

Sesquiterpene cyclases utilizing farnesyl pyrophosphate have been
isolated from plant tissues, most notably the enzyme from sage, *Salvia
officinalis* (Labiatae), involved in the biosynthesis of humulene and (−)-β-
caryophyllene (Croteau and Gundy, 1984). Two monoterpene cyclases
have also been isolated from sage leaves (Gambliel and Croteau, 1984),
catalysing different stereospecific cyclizations of geranyl pyrophosphate.
All these enzymes were found in soluble, membrane-free extracts and
required divalent metal ions for activity. It may be that the cyclase that
produces drimenol is very much less stable than those from sage, or is
present in considerably smaller amounts. The conversion of drimenol into
polygodial by *P. hydropiper in vitro* has been demonstrated (D.L.
Hallahan, unpublished), confirming the subsequent steps in the pathway,
but it is possible that drimenol is not synthesized by the proposed route.
Further work is needed to confirm the biosynthetic pathway and identify
the enzymes involved.

In theory, these terpenoid antifeedants are attractive for biotechnological transfer to other species, as the putative precursors are ubiquitous. The
small number of enzymic steps apparently involved in polygodial synthesis
from farnesyl pyrophospate should make transfer of the pathway relatively
simple. However, the inability to demonstrate cell-free conversion of
farnesyl pyrophosphate to drimenol is a major stumbling-block, exacer-
bated by the very low rates of polygodial synthesis by any easily grown bulk
tissue. One possible way around these difficulties might be a 'shotgun'
cloning of *P. hydropiper* or *G. odoratum* genes into a suitable host,

followed by screening for drimenol synthesis. In this way, the gene for any enzyme involved in drimenol synthesis could be identified without going via the protein. This approach does, however, rely on 'drimenol cyclase' being the product of a single, or two closely linked, gene(s).

2.3 Iridoid sex pheromones

(4a*R*,7*S*,7a*S*)-Nepetalactone (Scheme 5) is a major secondary component of the labiate *Nepeta mussinii* (about 0.1% of fresh weight). It can be easily converted by reduction to the nepetalactols XXII and XXIII, the sex pheromone of the hop aphid, *Phorodon humuli* (Dawson *et al.*, 1989b; Campbell *et al.*, 1990). In the plant, nepetalactone is presumably synthesized from geraniol pyrophosphate, through geraniol via 10-hydroxy-geraniol, oxidation to 10-oxogeranial, and cyclization (Inouye and Uesato, 1986). This pathway (Scheme 5) is common to the synthesis of a range of iridoid monoterpenes and the biochemistry is relatively well understood.

The first step in the pathway is catalysed by geraniol hydroxylase, and this has now been shown to be a cytochrome P450-linked enzyme in *N. mussinii*. Microsomal membranes enriched in cytochrome P450 monoterpene hydroxylase activity have been extracted from *Catharanthus roseus* (Apocynaceae) (Madyastha *et al.*, 1976)) and *N. mussinii* (D.L. Hallahan, unpublished). The former cytochrome P450 enzyme has recently been purified (Meijer *et al.*, 1990), although no sequence information is yet available. Sequence analysis of ripening-related cDNAs from avocado, *Persea americana* (Lauraceae), fruit have, however, yielded the first ever sequence of a plant cytochrome P450 (Bozak *et al.*, 1990). Synthetic oligonucleotide probes, based on conserved regions of cytochrome P450 sequences, are currently being employed to isolate P450 genes from *N. mussinii*, using polymerase chain reaction (PCR) techniques.

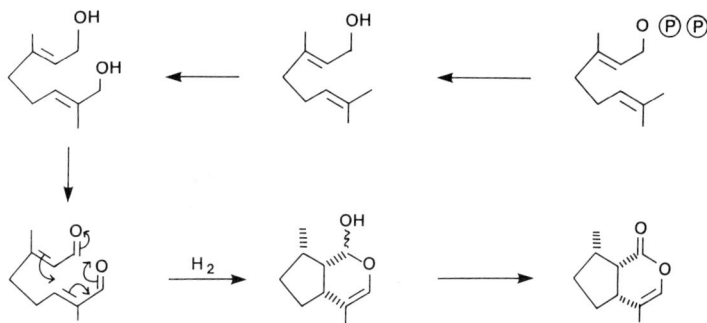

Scheme 5

Subsequent steps in the pathway have been studied in extracts of
Rauwolfia serpentina (Apocynaceae). In the presence of NAD, 10-
hydroxygeraniol is converted into 10-oxogeranial and a mixture of the
mono-oxidized intermediates. The additional presence of NADH leads to
the cyclization of 10-oxogeranial to iridodial (Uesato *et al.*, 1986, 1987).
This reductive cyclization is rather unusual, contrasting with the pyro-
phosphate-dependent cyclases more commonly associated with terpene
biosynthesis. The NAD(P)H-dependent oxogeranial cyclase was purified
440-fold from *R. serpentina* cell suspension cultures, and appeared to be a
homotetramer of 28 700 molecular weight subunits (Uesato *et al.*, 1987).

3. Future strategies

3.1 Targeting and control of defence gene expression

The development of genetic manipulation techniques for crop protection
may be seen as occurring in two phases. The initial phase, examples of
which have been given in other chapters of this volume (Chapter 7),
consists of the transfer of genes encoding single polypeptide products (e.g.
peptide protease inhibitors) under constitutive promoter control. Advances
with this approach have been rapid (see Chapter 2) and have led to the
successful transformation of at least 30 different plant species (Gasser and
Frayley, 1989). The use of the *Agrobacterium* nopaline synthase (*nos*)
promoter and the cauliflower mosaic virus 35S (*CaMV*) promoter has
enabled constitutive expression of novel genes in several plant species.
Transformants thus produced have been easily screened for successful
expression by incorporation of antibiotic-resistance genes in the T-DNA,
or by using antibodies to the polypeptide gene product.
   The second phase will, however, require additional complexity both in
conceptual approach and technical ability, as it will most likely involve:

1. The transfer of biosynthetic pathways involving multiple enzymes (e.g.
antifeedant biosynthetic pathways).
2. Control of the expression of genes both spatially (e.g. systemic v.
localized expression) and temporally (e.g. using promoters giving faster
responses than found in the genetic source plant).
3. Integration of the transferred genes with the host plant defence
response.

   Whilst the ability to transform plants to express multi-enzyme pathways
is accessible to present technology, it has yet to be achieved. Prosen and
Simpson (1987) reported that transfer of a model library of 10 members
into tobacco, *Nicotiana tabacum* (Solanaceae), using the *Agrobacterium*

system, resulted in a maximum incorporation of four distinct T-DNAs in a single regenerated plant. In addition, we still lack the ability to transform cereal crops routinely, though the use of microprojectile techniques may well overcome this problem in the near future (Gasser and Frayley, 1989).

The central problem facing the second-phase use of genetic manipulation technology will be in the *control* of gene expression. 'Yield penalties' are often described as potentially limiting the commercial viability of transformed crops, so the ability to switch on defence genes only when required would be perceived as a major advance. Where the production of potentially toxic or poor-tasting defence compounds is involved, localisation of expression would be essential to their usefulness as pest control agents in crop plants. Genes which exhibit organ-specific expression have been identified (Gasser and Frayley, 1989), regulatory sequences from which will enable precise targeting of gene expression to specific tissues within transformed plants.

Regulation of plant gene expression is an area which has received a great deal of attention (Kuhlemeier *et al.*, 1987), and we have perhaps reached a point where sufficient advances have been made to begin to integrate these discoveries into genetic manipulation programmes for pest control.

Plants respond to a number of environmental changes by specific gene activation. Light-induced genes have been identified, e.g. those encoding the small subunit of ribulose-1,5-bisphosphate carboxylase (*rbcS*) or the chlorophyll a/b binding protein (*cab*), as well as 'heat-shock' and drought-response proteins (Kuhlemeier *et al.*, 1987; Gomez *et al.*, 1988; Guerrero and Mullet, 1988). Plants also respond to pest and pathogen attack, generally by a broad-spectrum defence response involving the *de novo* production of phytoalexins, PR proteins (hydroxyproline-rich glycoproteins, glucanases, chitinases) and protease inhibitors, some of which are also produced in response to wounding (Green and Ryan, 1972; Chen and Varner, 1985; Lamb *et al.*, 1987). The control mechanisms of this response have received much attention. Albersheim *et al.* (1983) and Ryan (1987) have proposed that plant cells respond to cell wall-derived oligosaccharides, produced locally by insect feeding or pathogen attack, which they termed 'oligosaccharins'. Such pectic fragments have been shown to act as elicitors of defence-response genes for phytoalexin biosynthesis (Darvill and Albersheim, 1984) and of anti-auxin hormones (York *et al.*, 1984; McDougall and Fry, 1988) and to affect the morphogenesis of plant cell cultures (Eberhard *et al.*, 1989). Their role as wound-response messengers in whole plants, however, depends crucially on their ability to be transported throughout the plant. Baydoun and Fry (1985) performed a definitive experiment to test this hypothesis, by administering radiolabelled pectic polysaccharides to wound sites on leaves of tomato, *Lycopersicon esculentum* (Solanaceae). They reported that fragments with degree of polymerization 6–14 were not mobile away from the injury site, and proposed that pectic fragments might act locally in the release of a second, long-distance, messenger, yet to be identified. It is, however, conceivable that fragments with degree of polymerization < 6 are in fact translocatable (Ryan, 1987). Two recent reports (Malamy *et al.*, 1990; Métraux *et al.*, 1990) have, however, suggested a role for salicylic acid as the endogenous signal responsible for induced viral resistance, thus this compound may constitute the long-distance messenger.

Whatever the nature of the messenger, the next step in the elicitation of the plant's defensive response must involve its interaction with a receptor, with the resulting induction of gene expression. Studies on the phenylpropanoid phytoalexin pathway enzymes, phenylalanine-ammonia lyase (PAL) and chalcone isomerase (CHI), show that they are rapidly induced by elicitors in cultured bean cells (transcription is induced within 5 min of administration of the elicitor). Thus, there would appear to be few intervening steps between messenger binding to the receptor and gene transcription (Lamb *et al.*, 1987). Lamb *et al.* (1987) proposed two mechanisms for this phase of induction: interaction of an elicitor/receptor complex with *cis*-acting regulatory sequences, or the involvement of a signal–transduction cascade leading to modulation of *trans*-acting regulatory proteins which interact with *cis*-acting regulatory sequences. The latter mechanism might involve cyclic AMP, protein phosphorylation or phosphatidylinositol turnover (Ranjeva and Boudet, 1987).

Several regulatory sequences acting on wound-induced genes have been identified which may be applicable in controlling the expression of defence genes in transformed plants. The wound-inducible genes examined thus far show different expression patterns, both in induction time and pattern of expression. Protease inhibitor II genes in potato show maximum induction after 24 h, and their expression is systemic (Pena-Cortes *et al.*, 1988). The TR'2 promoter shows local expression (i.e. at the wound site) (Teeri *et al.*, 1989), as does a promoter of the glycine-rich cell wall protein GRP 1.8 (Keller *et al.*, 1989), with expression maximal in this case 30–45 min after wounding. Genes controlling the expression of the phenylpropanoid pathway enzymes PAL, CHI and CHS exhibit elevated transcription after 5 min, with maximal level of expression within 3–4 h (Lamb *et al.*, 1987). The spatial and temporal patterns of expression of the wound-inducible gene *win2* from potato (whose gene product is unknown), fused with β-glucuronidase (GUS) in transgenic potato and tobacco plants, have been studied (Stanford *et al.*, 1989, 1990). Wound induction of GUS activity was observed in potato transformants but not in the heterologous host, tobacco. Expression in wounded leaves and stems increased over a time-scale of six days, with GUS activity eventually extending beyond the vascular system until it could be detected throughout wounded leaves. It was therefore proposed that the wound stimulus was transported by the plant's vascular system. Apart from such *in vitro* studies, the expression of chloramphenicol acetyl transferase (CAT) protein, controlled by the wound-inducible proteinase inhibitor II (*pin2*) gene, has been examined in transformed tobacco under field conditions (Thornburg *et al.*, 1990). It was found that attack by phytophagous insects resulted in the expression of CAT activity, initially in the wounded parts of the leaf but, after 48 h, throughout the plant. Interestingly, it was found that significant expression resulted from attack by insects with chewing mouthparts, but that aphid infestation did not induce CAT activity. The work, however, demonstrates the feasibility of using wound-inducible promoters to control defence gene expression in transformed crop plants under natural conditions. Sequences upstream of the transcription start site of *win2* and *pin2* were shown by deletion analysis to be necessary for wound inducibility (Keil *et al.*, 1990; Stanford *et al.*, 1990), and further characterization of such promoter sequences should facilitate the design of tailored promoters for the expression of defence genes in transformed plants.

The use of such defined promoter sequences, in conjunction with genes for pest control agents, may thus give control over the temporal and spatial expression of defence genes in transformed crop plants in the near future. This would allow the production of crops with novel defence genes, induced only when required as part of the plant's response to pest or pathogen attack. To achieve these aims, the basic biochemistry and molecular biology of the control of plant gene expression must be understood. In addition, parallel investigations into insect–plant interactions will be required; a recent report describes the removal of partially eaten leaves of *Eucalyptus* by a number of insect species (Edwards and Wanjura, 1989), and the authors speculate that such behaviour may constitute 'sabotage' of the plant's defences by preventing the transmission of compound(s) that induce such responses.

Conclusions

As has been seen, there is significant scope for useful manipulation of secondary metabolism in crop plants. Although further chemical ecology studies are essential, a number of targets for enzymic and genetic studies can be identified. Some idea of the considerable problems involved in studying the enzymes of secondary metabolite biosynthesis should be apparent. The important role of cytochrome P450-dependent enzymes in the biosynthesis of secondary metabolites is becoming clear and recent advances in the purification, sequencing and cloning of such proteins is encouraging for the manipulation of these pathways.

References

Albersheim, P., Darvill, A.G., McNeil, M., Valent, B.S., Sharp, J.K., Nothnagel, E.A., Davis, K.R., Yamazaki, N., Gollin, D.J., York, W.S., Dudman, W.F., Darvill, J.E. and Dell, A. (1983) Oligosaccharins: naturally occurring carbohydrates with biological regulatory functions. In *Structure and Function of Plant Genomes*, Vol. III, ed. O. Cliferi and L. Dure. Plenum, New York, pp. 293–312.

Ames, B.N. and Gold, L.S. (1990) Misconceptions on pollution and the causes of cancer. *Angewandte Chemie* **29**, 1197–1208.

Ames, B.N., Profet, M. and Gold, L.S. (1990) Dietary pesticides (99.99% all natural). *Proceedings of the National Academy of Sciences (USA)* **87**, 7777–7781.

Asakawa, Y., Dawson, G.W., Griffiths, D.C., Lallemand, J.-Y., Ley, S.V., Mori, K., Mudd, A., Pezechk-Leclaire, M., Pickett, J.A., Watanabe, H., Woodcock, C.M. and Zhang, Z.-N. (1988) Activity of drimane antifeedants and related compounds against aphids, and comparative biological effects and chemical reactivity of (−)- and (+)-polygodial. *Journal of Chemical Ecology* **14**, 1845–1855.

Bailey, J.A. (1982) Mechanism of phytoalexin accumulation. In *Phytoalexins*, ed. J.A. Bailey and J.W. Mansfield, Blackie, Glasgow and London, pp. 289–323.

Banthorpe, D.V., Brooks, C.J.W., Brown, J.T., Lappin, G.J. and Morris, G.S. (1989) Synthesis and accumulation of polygodial by tissue cultures of *Polygonum hydropiper*. *Phytochemistry* **28**, 1631–1633.

Báthori, M., Máthé, I. Jr., Solymosi, P. and Szendrei, K. (1987) Phytoecdysteroids in some species of Caryophyllaceae and Chenopodiaceae. *Acta Botanica Hungarica* **33**, 377–385.

Baydoun, E.A.-H. and Fry, S.C. (1985) The immobility of pectic substances in injured tomato leaves and its bearing on the identity of the wound hormone. *Planta* **165**, 269–276.

Bestmann, H.-J., Classen, B., Kobold, U., Vostrowsky, O., Klingauf, F. and Stein, U. (1988) Steam volatile constituents from leaves of *Rhus typhina*. *Phytochemistry* **27**, 85–90.

Birch, A.N.E., Griffiths, D.W. and Smith, W.H.M. (1990) Changes in forage and oilseed rape root glucosinolates in response to attack by turnip root fly (*Delia floralis*). *Journal of the Science of Food and Agriculture* **51**, 309–320.

Bjostad, L.B. and Roelofs, W.L. (1983) Sex pheromone biosynthesis in *Trichoplusia ni:* key steps involve delta-11 desaturation and chain-shortening. *Science* **220**, 1387–1389.

Blight, M.M., Pickett, J.A., Smith, M.C. and Wadhams, L.J. (1984) An aggregation pheromone of *Sitona lineatus*. *Naturwissenschaften* **71**, 480–481.

Blight, M.M., Pickett, J.A., Wadhams, L.J. and Woodcock, C.M. (1989) Antennal response of *Ceutorhynchus assimilis* and *Psylliodes chrysocephala* to volatiles from oilseed rape. *Aspects of Applied Biology* **23**, 329–334.

Borg-Karlson, A.-K. (1987) Chemical basis for the relationship between *Ophrys* orchids and their pollinators III. *Chemica Scripta* **27**, 313–325.

Borg-Karlson, A.-K., Bergström, G. and Kullenberg, B. (1987) Chemical basis for the relationship between *Ophrys* orchids and their pollinators II. *Chemica Scripta* 27, 303–311.

Bowers, W.S. (1985) Phytochemical disruption of insect development and behavior. In *Bioregulators for Pest Control*, ed. P.A. Hedin. ACS Symposium Series 276, American Chemical Society, Washington, DC, pp. 225–236.

Bowers, W.S. (1991) Insect hormones and antihormones in plants. In *Herbivores: Their Interaction with Secondary Plant Metabolites* ed. G.A. Rosenthal and M.R. Berenbaum. Harcourt Brace Jovanovich, San Diego, in press.

Bozak, K.R., Yu, H., Sirevag, R. and Christoffersen, R.E. (1990) Sequence analysis of ripening-related cytochrome P450 cDNAs from avocado fruit. *Proceedings of the National Academy of Sciences (USA)* 87, 3904–3908.

Broughton, H.B., Ley, S.V., Slawin, A.M.Z., Williams, D.J. and Morgan, E.D. (1986) X-ray crystallographic structure determination of detigloyldihydro-azadirachtin and reassignment of the structure of the limonoid insect anti-feedant azadirachtin. *Chemical Communications* 1, 46–47.

Campbell, C.A.M., Dawson, G.W., Griffiths, D.C., Pettersson, J., Pickett, J.A., Wadhams, L.J. and Woodcock, C.M. (1990) The sex attractant pheromone of the damson–hop aphid *Phorodon humuli* (Homoptera, Aphididae). *Journal of Chemical Ecology* 16, 3455–3465.

Camps, F. (1991) Plant ecdysteroids and their interactions with insects. In *Annual Proceedings of the Phytochemical Society of Europe*, ed. J.B. Harborne and F.A. Tomas-Barberan. Clarendon Press, Oxford, pp. 331–376.

Casida, J.E. (ed.) (1973) *Pyrethrum: The Natural Insecticide.* Academic Press, New York and London.

Chapple, C.C.S., Glover, J.R. and Ellis, B.E. (1990) Purification and characterisa-tion of methionine:glyoxylate aminotransferase from *Brassica carinata* and *Brassica napus. Plant Physiology* 94, 1887–1896.

Chen, J. and Varner, J.E. (1985) An extracellular matrix protein in plants: characterization of a genomic clone for carrot extension. *EMBO Journal* 4, 2145–2151.

Critchley, B.R., Campion, D.G., McVeigh, L.J., McVeigh, E.M., Cavanagh, G.G., Hosny, M.M., Nasr, E.-S.A., Khidr, A.A. and Naguib, M. (1985) Control of pink bollworm, *Pectinophora gossypiella* (Saunders) (Lepidoptera: Gelechii-dae), in Egypt by mating disruption using hollowfibre, laminate-flake and microencapsulated formulations of synthetic pheromone. *Bulletin of Entomo-logical Research* 75, 329–345.

Croteau, R. and Gundy, A. (1984) Cyclisation of farnesyl pyrophosphate to the sesquiterpene olefins humulene and caryophyllene by an enzyme system from sage. *Archives of Biochemistry and Biophysics* 233, 838–841.

Darvill, A.G. and Albersheim, P. (1984) Phytoalexins and their elicitors: a defence against microbial infection in plants. *Annual Review of Plant Physiology* 35, 243–275.

Da Silva, M.L., Luz, A.I.R., Zoghbi, M.G.B., Ramos, L.S. and Maia, J.G.S. (1984) Essential oils of some Amazonian *Mikania* species. *Phytochemistry* 23, 2374–2376.

Dawson, G.W., Griffiths, D.C., Pickett, J.A., Smith, M.C. and Woodcock, C.M. (1984) Natural inhibition of the aphid alarm pheromone. *Entomologia*

Experimentalis et Applicata **36**, 197–199.

Dawson, G.W., Griffiths, D.C., Pickett, J.A., Wadhams, L.J. and Woodcock, C.M. (1987) Plant-derived synergists of the alarm pheromone from the turnip aphid, *Lipaphis (Hyadaphis) erysimi* (HOMOPTERA, APHIDIDAE). *Journal of Chemical Ecology* **13**, 1663–1671.

Dawson, G.W., Hallahan, D.L., Mudd, A., Patel, M.M., Pickett, J.A., Wadhams, L.J. and Wallsgrove, R.M. (1989a) Secondary plant metabolites as targets for genetic modification of crop plants for pest resistance. *Pesticide Science* **27**, 191–201.

Dawson, G.W., Janes, N.F., Mudd, A., Pickett, J.A., Slawin, A.M.Z., Wadhams, L.J. and Williams, D.J. (1989b) The aphid sex pheromone. *Pure and Applied Chemistry* **61**, 555–558.

Dawson, G.W., Griffiths, D.C., Merritt, L.A., Mudd, A., Pickett, J.A., Wadhams, L.J. and Woodcock, C.M. (1990) Aphid semiochemicals – a review, and recent advances on the sex pheromone. *Journal of Chemical Ecology* **16**, 3019–3030.

Deverall, B.J. (1982) Introduction. In *Phytoalexins*, ed. J.A. Bailey and J.W. Mansfield. Blackie, Glasgow and London, pp. 1–20.

Dickens, J.C. (1989) Green leaf volatiles enhance aggregation pheromone of boll weevil, *Anthonomus grandis. Entomologia Experimentalis et Applicata* **52**, 191–203.

Dickens, J.C., Jang, E.B., Light, D.M. and Alford, A.R. (1990) Enhancement of insect pheromone responses by green leaf volatiles. *Naturwissenschaften* **77**, 29–31.

Dobson, H.E.M., Bergström, J., Bergström, G. and Groth, I. (1987) Pollen and flower volatiles in two *Rosa* species. *Phytochemistry* **26**, 3171–3173.

Domon, B. and Hostettman, K. (1984) New saponins from *Phytolacca dodecandra l'HERIT. Helvetica Chimica Acta* **67**, 1310–1315.

Doughty, K.J., Porter, A.J.R., Morton, A.M., Kiddle, G., Bock, C.H. and Wallsgrove, R.M. (1991) Variation in the glucosinolate content of oilseed rape leaves. II. Response to infection by *Alternaria brassicae* (Berk.) Sacc. *Annals of Applied Biology* **118**, 467–477.

Downum, K.R. (1986) Photoactivated biocides from higher plants. In *Natural Resistance of Plants to Pests*, ed. M.B. Green and P.A. Hedin. ACS Symposium Series 296, American Chemical Society, Washington, DC, pp. 197–205.

Eberhard, S., Doubrava, N., Marfa, V., Mohnen, D., Southwick, A., Darvill, A. and Albersheim, P. (1989) Pectic cell wall fragments regulate tobacco thin-cell-layer explant morphogenesis. *Plant Cell* **1**, 747–755.

Edwards, P.B. and Wanjura, W.J. (1989) Eucalypt-feeding insects bite off more than they can chew: sabotage of induced defences? *Oikos* **54**, 246–248.

Elliott, M. (1979) Progress in the design of insecticides. *Chemistry and Industry*, 757–768.

Elliott, M. (1990) The contribution of pyrethroid insecticides to human welfare. *Actualité chimique*, 57–70.

Fenwick, G.R., Heaney, R.K. and Mullin, W.J. (1983) Glucosinolates and their breakdown products in food and food plants. *Critical Reviews in Food Science and Nutrition* **18**, 123–201.

Friend, J. (1985) Phenolic substances and plant disease. In *The Biochemistry of*

Plant Phenolics: Annual Proceedings of the Phytochemical Society of Europe, Vol. 25, ed. C.F. Van Sumere and P.J. Lea. Clarendon Press, Oxford, pp. 367–392.

Friend, J. and Threlfall, D.R. (eds) (1976) *Biochemical Aspects of Plant–Parasite Relationships: Annual Proceedings of the Phytochemical Society,* Vol. 13. Academic Press, London, New York and San Francisco.

Gambliel, H. and Croteau, R. (1984) Pinene cyclases I and II. *Journal of Biological Chemistry* **259**, 740–748.

Gasser, C.S. and Fraley, R.T. (1989) Genetically engineering plants for crop improvement. *Science* **244**, 1293–1299.

Gershenzon, J., Rossiter, M., Mabry, T.J., Rogers, C.E., Blust, M.H. and Hopkins, T.L. (1985) Insect antifeedant terpenoids in wild sunflower: a possible source of resistance to the sunflower moth. In *Bioregulators for Pest Control,* ed. P.A. Hedin. ACS Symposium Series 276, American Chemical Society, Washington, DC, pp. 433–446.

Glen, D.M., Jones, H. and Fieldsend, J.K. (1989) Effect of glucosinolates on slug damage to oilseed rape. *Aspects of Applied Biology* **23**, 377–381.

Glendening, T.M. and Poulton, J.E. (1990) Partial purification and characterisation of a 3′-phosphoadenosine-5′-phosphosulphate : desulphoglucosinolatesulfo- transferase from cress (*Lepidium sativum*). *Plant Physiology* **94**, 811–818.

Gomez, J., Sanchez-Martinez, D., Steifel, V., Rigau, J., Puigdomenech, P. and Pages, M. (1988) A gene induced by the plant hormone abscissic acid in response to water stress encodes a glycine-rich protein. *Nature* **334**, 262–264.

Goodwin, T.W. and Mercer, E.I. (eds) (1983) *Introduction to Plant Biochemistry,* 2nd edn. Pergamon Press, Oxford and New York.

Green, M.B. and Hedin, P.A. (eds) (1986) *Natural Resistance of Plants to Pests.* ACS Symposium Series 296, American Chemical Society, Washington, DC.

Green, T.R. and Ryan, C.A. (1972) Wound-induced proteinase inhibitor in plant leaves: a possible defence mechanism against insects. *Science* **175**, 776–777.

Guerin, P.M., Städler, E. and Buser, H.R. (1983) Identification of host plant attractants for the carrot fly, *Psila rosae. Journal of Chemical Ecology* **9**, 843–861.

Guerrero, D.F. and Mullet, J.E. (1988) Reduction in turgor induces rapid changes in leaf translatable RNA. *Plant Physiology* **88**, 401–408.

Halkier, B.A., Scheller, H.V. and Lindberg Moller, B. (1989) The involvement of cytochrome P450 in the biosynthesis of cyanogenic glucosides. In *Cytochrome P-450: Biochemistry and Biophysics,* ed. I. Schuster. Taylor and Francis, London and New York, pp. 154–157.

Hanssen, H.P. and Abraham, W.R. (1987) Odiferous compounds from liquid culture of *Gloeophyllum odoratum* and *Lentillinus cochleatus* (Basidiomyco- tina). *Flavour and Fragrance Journal* **2**, 171–174.

Harborne, J.B. (1985) Phenolics and plant defence. In *The Biochemistry of Plant Phenolics: Annual Proceedings of the Phytochemical Society of Europe,* Vol. 25, ed. C.F. Van Sumere and P.J. Lea. Clarendon Press, Oxford, pp. 393–408.

Hartmann, T., Ehmke, A., Eilert, U., von Borstel, K. and Theuring, C. (1989) Sites of synthesis, translocation and accumulation of pyrrolizidine alkaloid N-oxides in *Senecio vulgaris* L. *Planta* **177**, 98–107.

Hartmann, T., Biller, A., Witte, L., Ernst, L. and Boppré, M. (1991) Transformation of plant pyrrolizidine alkaloids into novel insect alkaloids by arctiid moths (Lepidoptera). *Biochemical Systematics and Ecology* (in press).

Helmlinger, J., Rausch, T. and Hilgenberg, W. (1985) Metabolism of ^{14}C-indole-3-acetaldoxime by hypocotyls of Chinese cabbage. *Phytochemistry* **24**, 2497–2502.

Hiller, K. (1987) New results on the structure and biological activity of triterpenoid saponins. In *Biologically Active Natural Products: Annual Proceedings of the Phytochemical Society of Europe*, Vol. 27, ed. K. Hostettmann and P.J. Lea. Clarendon Press, Oxford, pp. 167–184.

Inouye, H. and Uesato, S. (1986) Biosynthesis of ridoids and secoiridoids. *Progress in the Chemistry of Organic Natural Products*, Vol. 50, ed. W. Herz, H. Grieseback, G.W. Kirby and C. Jamm. Springer-Verlag, New York, pp. 169–236.

Jacobson, M. (1990) *Glossary of Plant-Derived Insect Deterrents*. CRC Press, Boca Raton, Florida.

Jain, J.C., Groot Wassink, J.W.D., Reed, D.W. and Underhill, E.W. (1990) Persistent co-purification of enzymes catalysing the sequential glucosylation and sulfation steps in glucosinolate biosynthesis. *Journal of Plant Physiology* **136**, 356–361.

Kami, T. (1975) Identification of components in the essential oil of Hybridsorgo, a forage sorghum. *Journal of Agricultural and Food Chemistry* **23**, 795–798.

Keil, M., Sanchez-Serrano, J., Schell, J. and Wilmitzer, L. (1990) Localization of elements important for the wound-inducible expression of a chimeric potato proteinase inhibitor II-CAT gene in transgenic tobacco plant. *Plant Cell* **2**, 61–70.

Keller, B., Schmid, J. and Lamb, C.J. (1989) Vascular expression of a bean cell wall glycine-rich protein – β-glucuronidase gene fusion in transgenic tobacco. *EMBO Journal* **8**, 1309–1314.

Kindl, H. and Underwood, E.W. (1968) Biosynthesis of mustard oil glucosides: N-hydroxyphenylalanine, a precursor of glucotropaeolin and a substrate for the enzymatic and nonenzymatic formation of phenylacetaldehyde oxime. *Phytochemistry* **7**, 745–756.

Koritsas, V.M., Lewis, J.A. and Fenwick, G.R. (1989) Accumulation of indole glucosinolates in *Psylliodes chrysocephala* L.-infested or -damaged tissues of oilseed rape. *Experientia* **45**, 493–495.

Kraus, W., Bokel, M., Klenk, A. and Pöhnl, H. (1985) The structure of azadirachtin and 22,23-dihydro-23β-methoxyazadirachtin. *Tetrahedron Letters* **26**, 6435–6438.

Kuhlemeier, C., Green, P.J. and Chua, N.-H. (1987) Regulation of gene expression in higher plants. *Annual Review of Plant Physiology* **38**, 221–257.

Lamb, C.J., Bell, J.N., Corbin, D.R., Lawton, M.A., Mehdy, M.C., Ryder, T.B., Sauer, N. and Walter, M.H. (1987) Activation of defense genes in response to elicitor and infection. In *UCLA Symposia on Molecular and Cellular Biology: Molecular Strategies for Crop Protection, New Series*, Vol. 48, ed. C.J. Arntzen and C. Ryan. Alan R. Liss, New York, pp. 49–58.

Ley, S.V. (1990) Synthesis and modification of azadirachtin and related anti-feedants. In *Recent Advances in the Chemistry of Insect Control II*, ed. L. Crombie, Special Publication No. 79, Royal Society of Chemistry, Cambridge, pp. 90–98.

Ludwig-Muller, J. and Hilgenberg, W. (1988) A plasma membrane-bound enzyme oxidises L-tryptophan to indole-3-acetaldoxime. *Physiologia Plantarum* **74**, 240–250.

Ludwig-Muller, J., Rausch, T., Lang, S. and Hilgenberg, W. (1990) Plasma membrane bound high pI peroxidase isozymes convert tryptophan to indole-3-acetaldoxime. *Phytochemistry* **29**, 1397–1400.

Luthy, B. and Matile, P. (1984) The mustard oil bomb: rectified analysis of the subcellular organisation of the myrosinase system. *Biochemie und Physiologie der Pflanzen* **179**, 5–12.

Macaulay, E.D.M., Etheridge, P., Garthwaite, D.G., Greenway, A.R., Wall, C. and Goodchild, R.E. (1985) Prediction of optimum spraying dates against pea moth, *Cydia nigricana* (F.), using pheromone traps and temperature measurements. *Crop Protection* **4**, 85–98.

McDougall, G.J. and Fry, S.C. (1988) Inhibition of auxin-stimulated growth of pea stem segments by a specific nonasaccharide of xyloglucan. *Planta* **175**, 412–416.

McFarlane, I.J., Lees, E.M. and Conn, E.E. (1975) The *in vitro* biosynthesis of dhurrin, the cyanogenic glycoside of *Sorghum bicolor. Journal of Biological Chemistry* **250**, 4708–4713.

Madyastha, K.M., Meehan, T.D. and Coscia, C.J. (1976) Characterisation of a cytochrome P450-dependent monoterpene hydroxylase from the higher plant *Vinca rosea. Biochemistry* **15**, 1097–1102.

Mahadevan, S. (1973) Role of oximes in nitrogen metabolism in plants. *Annual Review of Plant Physiology* **24**, 69–88.

Malamy, J., Carr, J.P., Kiessig, D.F. and Raskin, I. (1990) Salicylic acid: a likely endogenous signal in the resistance response of tobacco to viral infection. *Science* **250**, 1002–1004.

Marston, A. and Hostettmann, K. (1991) Plant saponins: chemistry and molluscicidal action. In *Annual Proceedings of the Phytochemical Society of Europe* ed. J.B. Harborne and F.A. Tomas-Barberan, Clarendon Press, Oxford, pp. 264–286.

Matthews, D.L., Michalak, P.S. and Macrae, R.J. (1983) The effect of traditional insect-repellent plants on insect numbers in a mixed planting system. In *Proceedings of the 4th International Federation of Organic Agricultural Movements Conference, Massachusetts Institute of Technology, Cambridge, Massachusetts, 18–20 August 1982.* ed. W. Lockeretz, Praeger, Boston, pp. 117–127.

Mauchamp, B. and Pickett, J.A. (1987) Juvenile hormone-like activity of (E)-β-farnesene derivatives. *Agronomie* **7**, 523–529.

Meijer, A.H., Pennings, E.J.M., DeWaal, A. and Verpoorte, R. (1990) Purification of cytochrome P450-dependent geraniol-10-hydroxylase from a cell suspension culture of *Catharanthus roseus.* In *Progress in Plant Cellular and Molecular Biology,* ed. H.J.J. Nijkamp, L.H.W. Van der Plas and J. Van Aartrijk. Kluwer, Dordrecht, pp. 769–774.

Metcalf, R.L., Metcalf, R.A. and Rhodes, A.M. (1980) Cucurbitacins as kairomones for diabroticite beetles. *Proceedings of the National Academy of Sciences (USA)* **77**, 3769–3772.

Métraux, J.P., Signer, H., Ryals, J., Ward, E., Wyss-Benz, M., Gaudin, J.,

Raschdorf, K., Schmid, E., Blum, W. and Inveradi, B. (1990) Increase in salicylic acid at the onset of systemic acquired resistance in cucumber. *Science* **250**, 1004–1006.

Milford, G.F.J., Porter, A.J.R., Fieldsend, J.K., Miller, C.A., Leach, J.E. and Williams, I.H. (1989) Glucosinolates in oilseed rape (*Brassica napus*) and the incidence of pollen beetles (*Meligethes aeneus*). *Annals of Applied Biology* **115**, 375–380.

Nahrstedt, A. (1987) Recent developments in chemistry, distribution and biology of the cyanogenic glycosides. In *Biologically Active Natural Products. Annual Proceedings of the Phytochemical Society of Europe*, Vol. 27, ed. K. Hostettmann and P.J. Lea. Clarendon Press, Oxford, pp. 213–234.

Nault, L.R. and Styer, W.E. (1972) Effects of sinigrin on host selection by aphids. *Entomologia Experimentalis et Applicata* **15**, 423–437.

Nishida, R. and Fukami, H. (1990) Sequestration of distasteful compounds by some pharmacophagous insects. *Journal of Chemical Ecology* **16**, 151–164.

Nishida, R., Fukami, H., Miyata, T. and Takeda, M. (1989) Clerodendrins: feeding stimulants for the adult turnip sawfly, *Athalia rosae ruficornis*, from *Clerodendron trichotomum* (Verbenaceae). *Agricultural and Biological Chemistry* **53**, 1641–1645.

Nottingham, S.F., Hardie, J., Dawson, G.W., Hick, A.J., Pickett, J.A., Wadhams, L.J. and Woodcock, C.M. (1991) Behavioural and electrophysiological responses of aphids to host and non-host plant volatiles. *Journal of Chemical Ecology* **17**, 1231–1242.

Pena-Cortes, H., Sanchez-Serrano, J., Rocha-Sosa, M. and Willmitzer, L. (1988) Systemic induction of proteinase-inhibitor II gene expression in potato plants by wounding. *Planta* **174**, 84–89.

Pickett, J.A., Dawson, G.W., Griffiths, D.C., Hassanali, A., Merritt, L.A., Mudd, A., Smith, M.C., Wadhams, L.J., Woodcock, C.M. and Zhang, Z.-N. (1987) Development of plant-derived antifeedants for crop protection. In *Pesticide Science and Biotechnology*, ed. R. Greenhalgh and T.R. Roberts. Blackwell Scientific Publications, Oxford, pp. 125–128.

Pickett, J.A., Wadhams, L.J. and Woodcock, C.M. (1991) New approaches to the development of semiochemicals for insect control. In *Proceedings, Conference on Insect Chemical Ecology, Tábor, Czechoslovakia, 12–18 August 1990* (in press).

Prestwich, G.D. (1990) Proinsecticides: metabolically activated toxicants. In *Safer Insecticides: Development and Use*, ed. E. Hodgson and R.J. Kuhr. Marcel Dekker, New York, pp. 281–335.

Prosen, D.E. and Simpson, R.B. (1987) Transfer of a ten-member genomic library to plants using *Agrobacterium tumefaciens*. *Bio/Technology* **5**, 966–971.

Ranjeva, R. and Boudet, A.M. (1987) Phosphorylation of proteins in plants: regulatory effects and potential involvement in stimulus/response coupling. *Annual Review of Plant Physiology* **38**, 73–93.

Rawlinson, C.J., Doughty, K.J., Bock, C.H., Church, V.J., Milford, G.F.J. and Fieldsend, J.K. (1989) Diseases and responses to disease and pest control on single- and double-low cultivators of winter oilseed rape. *Aspects of Applied Biology* **23**, 393–400.

Rossiter, J.T., James, D.C. and Atkins, N. (1990) Biosynthesis of 2-hydroxy-3-

butenylglucosinolate and 3-butenylglucosinolate in *Brassica napus. Phytochemistry* **29**, 2509–2512.

Ryan, C.A. (1987) Oligosaccharide signalling in plants. *Annual Review of Cell Biology* **3**, 295–317.

Shorey, H.H. and McKelvey, J.J. (eds) (1977) *Chemical Control of Insect Behaviour.* Wiley and Sons, New York and London.

Spencer, K.C. (ed.) (1988) *Chemical Mediation of Coevolution.* Academic Press, London, New York and San Diego.

Stanford, A., Bevan, M.W. and Northcote, D.H. (1989) Differential expression within a family of novel wound-induced genes in potato. *Molecular and General Genetics* **215**, 200–208.

Stanford, A.C. Northcote, D.H. and Bevan, M.W. (1990) Spatial and temporal patterns of transcription of a wound-induced gene in potato. *EMBO Journal* **9**, 593–603.

Teeri, T.H., Lehvaslaiho, H., Franck, M., Uotila, J., Heino, P., Palva, E.T., VanMontagu, M. and Herrera-Estrella, L. (1989) Gene fusions to *LacZ* reveal new expression patterns of chimeric genes in transgenic plants. *EMBO Journal* **8**, 343–350.

Thornburg, R.W., Kernan, A. and Molin, L. (1990) Cloramphenicol acetyl transferase (CAT) protein is expressed in transgenic tobacco in field tests following attack by insects. *Plant Physiology* **92**, 500–505.

Toong, Y.C., Schooley, D.A. and Baker, F.C. (1988) Isolation of insect juvenile hormone III from a plant. *Nature* **333**, 170–171.

Uesato, S., Ogawa, Y., Inouye, H., Saiki, K. and Zenk, M.H. (1986) Synthesis of iridodial by cell free extracts from *Rauwolfia serpentina* cell suspension cultures. *Tetrahedron Letters* **27**, 2893–2896.

Uesato, S., Ikeda, H., Fujita, T., Inouye, H. and Zenk, M.H. (1987) Elucidation of iridodial formation mechanism – partial purification and characterisation of the novel monoterpene cyclase from *Rauwolfia serpentina* cell suspension cultures. *Tetrahedron Letters* **28**, 4431–4434.

Vaughan, J.G., MacLeod, A.J. and Jones, B.M.G. (eds) (1976) *The Biology and Chemistry of the Cruciferae.* Academic Press, London, New York and San Francisco.

Visser, J.H. (1983) Differential sensory perceptions of plant compounds by insects. In *Plant Resistance to Insects*, ed. P.A. Hedin. ACS Symposium Series **208**, American Chemical Society, Washington, DC, pp. 215–230.

Visser, J.H. (1986) Host odour perception in phytophagous insects. *Annual Review of Entomology* **31**, 121–144.

Visser, J.H. and Avé, D.A. (1978) General green leaf volatiles in the olfactory orientation of the Colorado beetle, *Leptinotarsa decemlineata. Entomologia Experimentalis et Applicata* **24**, 738–749.

Waller, G.R. (ed.) (1987) *Allelochemicals: Role in Agriculture and Forestry.* ACS Symposium Series 330, American Chemical Society, Washington, DC.

Wensler, R.J.D. (1962) Mode of host selection by an aphid. *Nature* **195**, 830–831.

Whitman, D.W. and Eller, F.J. (1990) Parasitic wasps orient to green leaf volatiles. *Chemoecology* **1**, 69–76.

York, W.S., Darvill, A.G. and Albersheim, P. (1984) Inhibition of 2,4-dichlorophenoxyacetic acid-stimulated elongation of pea stem segments by a xyloglucan oligosaccharide. *Plant Physiology* **75**, 295–297.

Zobel, A.M. and Brown, S.A. (1990) Dermatitis-inducing furanocoumarins on leaf surfaces of eight species of rutaceous and umbelliferous plants. *Journal of Chemical Ecology* **16**, 693–700.

Chapter 10
Promoting Crop Protection by Genetic Engineering and Conventional Plant Breeding: Problems and Prospects

Harold W.Woolhouse

Waite Agricultural Research Institute, University of Adelaide,
Glen Osmond, South Australia 5064

Careful analysis of well-preserved fossil deposits such as the Rhynie chert indicate that even in the early period of the land flora some 400 million years ago the co-evolution of insects and fungi with their plant hosts was under way. As the flora diversified so the range of predators evolved alongside it and there emerged new phyla such as the angiosperms and with them the herbivorous mammals. It is not surprising that in consequence of this very long period of co-evolutionary history the range of defence mechanisms which have evolved in plants is of enormous complexity. They include:

1. Defences against other competing plants: includes allelopathy, relative fecundity, timing mechanisms such as early germination and morphological adaptations such as shading and twining.

2. Defences against herbivory. The range of plant defences against herbivorous predators, both vertebrate and invertebrate, ranges from morphological features such as spines, thorns and thick cuticles to physiological mechanisms such as timing of appearance, capacity for wound response and rapid regrowth, and chemical defences, as for example enzymes to digest the invader and protease inhibitors which block its digestive system. There is also a vast array of secondary metabolites which serve to deter or poison the potential consumer. It is noteworthy that there are even examples of mimicry; examples of insect and reptile mimicry of toxic species are often popularized in scientific writings but many analogous examples are known in the plant kingdom.

3. Defences against fungi and bacteria. The narrow host range of many fungal and bacterial pathogens of plants testifies to the genetic specificity of the host–pathogen interaction. Antifungal and bacterial defences may be morphological as with extent of cuticle development, physiological as with necrotrophic and biotrophic responses of the host or biochemical as manifest in the vast array of cutinases and other enzyme responses, phytoalexins, and secondary metabolites which have fungicidal and bacteriocidal action.

4. Defences against viruses. The host range of plant viruses is very varied implying different degrees of specificity in the interaction of these molecular parasites with their hosts. Although in this case we know more about the nature of the pathogen, in some cases amounting to a complete knowledge of the nucleotide sequences in the virus genome and the crystal structure of its protein envelope, it is salutary to consider how little is known as yet of the nature of host plant responses and the nature of defences at the level of either the transmitting vectors or the limitation of cell-to-cell spread and replication within the host.

5. Over and above all of these mechanisms there exists an ecological level of adaptation of host and pathogen in which the frequency and spacing of the host plant on the ground may serve to combat the dispersive powers of the competitor or predator and restrict its capacity for the build-up of these dense populations.

I have left this ecological class of defences until last since it provides a convenient lead into an assessment of the scope for biotechnological contributions to crop protection. In the course of man's farming history we have seen the move to monocultures and thence through the agency of propagators and plant breeders to the narrowing of the genetic diversity in specific crops and the planting of genetically uniform material in monocultures over vast areas. The requirements of mechanical harvesting and very specific market requirements in both appearance and timing of crops have provided selection pressures to iron out everything from physiological and morphological diversity to flavour.

This brief survey serves to emphasize several things:

1. The enormous diversity of existing mechanisms of host–predator interaction, that is, the rich store in which the biotechnologist might choose to operate.

2. The extreme patchiness of our knowledge of even the physiological and biochemical nature of the interactions which the genetic engineer might seek to exploit, much less their molecular genetics.

3. That unless the efforts of genetic engineers and plant breeders can be matched by an equivalent leap forwards in the sophistication of agronomists and farmers the biotechnology will be of little significance.

I propose to illustrate these propositions by recourse to examples.

1. Insect control

Biotechnology in the service of pest control is undoubtedly gaining ground. Just a small sample of the methods being tried are: (i) insect baculoviruses as insecticides; (ii) inhibitors of digestive enzymes, including protease inhibitors (see Chapter 7); (iii) *Bacillus thuringiensis* toxic proteins (Chapter 6); (iv) Antifeedants against insect attack; (v) insect pheromones for luring pests to baited traps; and (vi) fungal insecticides.

One must emphasize that such is the long co-evolutionary history of insects with their hosts and their pathogens that these calculated designs for interfering with these relationships represent only a minute fraction of the possibilities which one may expect to be uncovered and exploited in the future.

It is important that the intrinsic optimism and excitement engendered by these new developments be kept in perspective. There is no doubt that in the more enlightened sectors of the farming community integrated pest management systems are gaining ground, but can we be confident of the foundations upon which this shift rests? For example, the move to use *B. thuringiensis* for protection of the maize crop against European corn borer (*Ostrinia nubilalis*) in the USA may seem laudable; it could even represent economic sense in times when the cost of oil is forcing up the price of manufactured crop protection chemicals and the emergence of insect resistance to insecticides is developing rapidly. There are undoubtedly social pressures to decrease use of chemical protectants as the insidious chronic effects of low-level exposure to many of the most potent compounds become documented and made public. Requirements of environmental legislation in states such as California are also working to discourage farmers from their former reliance on chemical controls. We should be uneasy about a headlong rush towards extremes in this situation. It seems certain that the heyday of the agrochemical industry has passed, but it is not over. Not all crop protection chemicals are harmful and economies from safer techniques of application are emerging based on lower dosage applications. One of the greatest problems of genetically engineered controls would seem to be that of resistance. If we take the case of *B. thuringiensis* toxin (see Chapter 6), progress is impressive: the protein can now be produced in bulk by fermentation at a cost which is competitive with chemical pesticides for high-value crops. Basic work on the protein has revealed two domains, one involved in binding and hence insect specificity, the other with toxicity, apparently through disruption of potassium channels in the lining of the gut of the insect which ingests the protein. The 'host range' of the toxic protein can be altered by manipulation of that portion of the gene which encodes the protein-binding domain. In transformable species such as potato and tomato, and most recently cotton, it has been possible to express the *B. thuringiensis* protein in the host plant. Initially levels of expression were insufficient to render the plants toxic to feeding insects but manipulation of codon usage by site-directed mutagenesis and improved promoters has largely overcome these problems. The central problem is that the appearance of vast acreages of genetically engineered plants containing a particular form of *B. thuringiensis* protein offers a massive sieve for the selection of mutant insects with resistance to the toxin. There are examples already of resistance to the protein used as an applied insecticide and selection will probably be more rapid when the protein is consistently present in the host.

The important point here is that these various forms of control which can be achieved by genetic engineering are based on interactions with components of the insect pest which are also subject to mutation and variation, and to that extent we are not in a new situation. It may be every bit as time-consuming and expensive to keep up with challenging the capacity of the pest to meet the selection pressures which the new materials place upon them.

2. Virus resistance

The development of an understanding of the molecular structure of plant viruses has opened up new avenues to the engineering of virus resistance. For example, inoculation of plants with virulent strains of some viruses in the presence of an excess of virus coat protein has been shown to afford some resistance to the virus. Extension of this finding to introducing coat protein genes from the virus to the host genome and thereby pre-arming the host with a supply of coat protein was a logical next step, and there are now reports of virus resistance being incorporated into a range of species by introduction of coat protein genes into the plant genome, using *Agrobacterium*-mediated transformation. With some viruses in other host species coat protein transfer has failed to work, but at present there is no clear understanding of how it works in successful cases, or of the basis for failure in others. The possibility exists to inhibit a plant virus at any point in its cycle of replication and spread, and along these avenues approaches to resistance are being made through studies of viral genes which regulate cell-to-cell spread, insect transmission and replication of the virus. In a few cases antisense RNA has been shown to be effective. Thus antisense to the 3′ portion of the tobacco mosaic virus genome, including the coat protein gene, resulted in low levels of protection against the virus; it seems that in this instance it is the presence of antisense sequences to the 3′ t-RNA-like terminal sequence which serves as a replicase-binding site which is the key to the effect.

Any reading of the recent progress on understanding the molecular structures of plant viruses and the development from this knowledge of new avenues of investigation of transmission, cell-to-cell spread and replication, is bound to give rise to a feeling of confidence that the conquest of many plant viruses may be imminent. We may be in for an unpleasant surprise. In the case of seed-propagated crops there arises the huge problem of stability of inheritance of the introduced genes, the question of levels of expression under a range of field conditions and the cost of doing the genetic transfer work. More important is the enormous intrinsic variability, particularly in the RNA viruses which comprise some 95% of the known plant viruses. In the case of perennial crops, introduction of genes offering protection may not require the expense of a breeding programme but the problems of clonal susceptibility in the face of mutation remain.

3. Herbicide resistance (see also Chapter 4)

Over the past decade we have witnessed the appearance in the market-place of crops showing increasing degrees of herbicide tolerance. The first examples to appear were mutants obtained by applying extreme selection pressures to the species. More recently crop plants carrying genes for resistance to such herbicides as bromoxymyl, atrazine and glyphosate have been produced. Finding resistant genotypes from within a crop using protoplast or tissue culture techniques affords the option of screening vast populations of cells for resistance to herbicides held to be environmentally benign and thereby increases the armoury of techniques available to farmers to couple with their agronomic practices. In this context it has to be emphasized that in some areas where arable agriculture is difficult, as for example the Mediterranean-type climatic zones, herbicides may be a key factor in sustaining minimum tillage practices which are crucial to conservation of the soil. There remains the problem that herbicide-tolerant crops will tend to increase herbicide use, thereby reducing biological diversity, that the technology of genetic engineering can be extended to environmentally harsh as well as benign herbicides, and that resistant crops may in some cases prove more efficient carriers of residues.

To the present author the balance of current evidence concerning the use of genetic engineering for herbicide resistance looks very dubious on environmental grounds but even more so when the economic framework of the enterprise is examined. If companies engineering herbicide-resistant plants are to achieve adequate returns on their investments, they will: (i) need to be operating in a framework within which the 'safety', i.e. acceptability, of the herbicide can be guaranteed over a long enough period to recoup the investment; (ii) need to be operating with crops in which the genotype can be protected from on-farm propagation by offering F_1 hybrid seed or clumsy legislative devices; (iii) have to take substantial risks concerning the long-term competitive position of the herbicide in question; and (iv) have to take account of the position of their chemical-based technology in the face of the biologically based sustainable systems which are now receiving intensive study. It could be argued that this represents a bleak prospect for genetic engineering of herbicide tolerance on both biological and economic grounds. In reality the position may be even worse than that which I have protrayed. There may be a place for glyphosate-resistance genes from bacteria expressed in trees, such as the poplars, now becoming available, but the gains will need to be carefully assessed, and even in cases such as this there are potential erosion problems in many areas arising from removal of the protective blanket of pioneer vegetation on potentially erodable sites. The catalogue of environmental concerns is rising as they relate to chronic effects which are poorly assessed: effects of surfactants, increased herbicide usage encouraged by tolerance in the crop leading to enhanced contamination of ground waters, and enhanced selection pressure favouring the emergence of herbicide tolerance in weed populations. To this catalogue there must be added the potential problems of the flow of resistance genes to related weedy species.

It is of interest to note that much of the effort in herbicide biotechnology is coming from the United States and Europe, where agricultural subsidies to the producer can buffer the system and justify the efforts of the chemical companies to produce resistant varieties. In these circumstances the benefits of the resistant crop go to the chemical company, not the primary producer, and it is argued by some that in many instances this simply accelerates the process of depopulation of rural areas.

4. Some ecological considerations

Work with cereals and potatoes has suggested that the sowing of varietal mixtures can lead to higher yields than are obtained from single varieties grown under the same conditions. The extent of these gains is often marginal and the biological basis of the phenomenon is not understood; it could result from physiological differences between genotypes enabling them to exploit different niches in the nutrient horizons of the soil, variation in host specificity limiting build-up of soil-borne pathogens, or a variety of other causes. In any event, these effects do point to the fact that biodiversity is likely to be an important key to successful sustainable agriculture in the future. In the examples that I have given of the emergence of genetic engineering in the realm of crop protection, there is immense novelty and imaginative thinking; what impact this will have on plant breeding is less clear. If one can isolate a set of resistance genes to a fungal pathogen in a cereal crop and insert them in parallel into a given variety of otherwise excellent quality, it may afford the option of keeping ahead of the pathogen by spatial and temporal heterogeneity in the planting of the lines with different resistance genes. It is evident, however, that this depends not only on isolation of the resistance genes and their successful expression in the chosen variety but also on the emergence of much more sophisticated sowing and agronomic arrangements in order to guarantee the mix of resistant crop genotypes from year to year, and place to place, in a given area. There are few parts of the world where such elaborate regulation of the agricultural industry will be really feasible.

I conclude where I begin with the concern that the co-evolution of plant hosts and their pests and pathogens has a long history, and we have but scratched the surface of the complexity of these interactions. It is my view that the present euphoria for biotechnology in respect of crop protection will be justified, but only in the long term. There may be some instances of relatively rapid success – engineered resistance to some viruses in some hosts may prove to be one of the better prospects. The greatest barrier to progress, however, will be our relative ignorance of the physiology, biochemistry, genetics and molecular biology of the vast majority of plant/pathogen interactions, primarily because research in the plant sciences over the past century has commanded only a fraction of the manpower and funding that has gone into research on animals. Until this background work is accomplished, the impact of biotechnology on plant breeding will be relatively modest and even when it is achieved it would seem likely that the most important developments will be in the application of genetic engineering to various aspects of the performance of biological control agents, rather than the simplistic approach of hoping to win the day by engineering changes in the genotype of the host alone.

5. The position of plant breeding

We have taken a brief and inevitably superficial look at some aspects of the scope for genetic manipulation in crop protection, but there remains the question of the impact this will have on plant breeding. The most obvious areas of application of this technology are in the introduction of markers as part of a breeder's RFLP (restriction fragment length polymorphism) mapping programme, the use of transposable elements to facilitate gene isolation and the use of cloning techniques such as yeast accessory chromosomes to assist with chromosome walking procedures for gene isolation. There will almost certainly be an acceleration of breeding programmes through such devices as the use of RFLP markers to facilitate selection for resistance in populations of thousands of seedlings without the need for the time, expense and trouble of growing vast numbers of plants to maturity. It is not clear, however, that these developments alone will guarantee anything approaching the full potential of genetic manipulation in plant breeding. To understand this we need to consider the realities of the position of the plant breeder.

Whether it be in a commercial company, a government research institute or a university department, the plant breeder tends to be a somewhat isolated figure. In the commercial environment confidentiality surrounds the operation, in government research there is often the pressure of industry levies on the breeder and in the academic environment resources are the most frequent problem. At the centre of the breeding operation, it is the performance of the commercial variety by which the breeders are judged. They must of necessity have the habits of magpies, gathering useful material from wherever they may, and there is also the pressure of the programme canalizing the efforts towards very specific objectives. In these circumstances many breeders enjoy the support of pathologists to devise their screening, some make use of cytogeneticists to help in areas such as wide crossing into the programme and a few make use of the work of physiologists in defining new characters for selection, but it is extremely rare to find breeding teams integrated to the level where crucial decisions in the programme come from the calculated deliberations of a close-knit team rather than the idiosyncratic activities of the individual breeder. If biotechnology is to have a real impact on plant breeding in a comprehensive way, then this well-tried, traditional situation will have to change. If we are to be introducing not single genes but groups of genes determining such characters as the biosynthetic pathway to a pest-resistance compound, or a thicker cuticle coupled with agronomically favourable traits, there will have to be team involvement and computer-assisted modelling to guide the breeding programme. If this level of teamwork in plant breeding does develop, then one can see long-term prospects for a major impact of biotechnology in plant breeding. I would conclude by emphasizing, however, that these efforts alone will be of little value unless there are parallel advances in agronomy and, if progress is achieved on both of these fronts, their impact will be set at nought unless means can be found to arrest the increase in the human population.

Index